T0291098

# THE WONDERS OF SCIENCE

A User Friendly Guide

# THE WONDERS
## OF SCIENCE
A User Friendly Guide

## Nathan Aviezer
Bar-Ilan University, Israel

World Scientific

EW JERSEY · LONDON · SINGAPORE · BEIJING · SHANGHAI · HONG KONG · TAIPEI · CHENNAI · TOKYO

*Published by*

World Scientific Publishing Co. Pte. Ltd.

5 Toh Tuck Link, Singapore 596224

*USA office:* 27 Warren Street, Suite 401-402, Hackensack, NJ 07601

*UK office:* 57 Shelton Street, Covent Garden, London WC2H 9HE

Library of Congress Control Number: 2024938769

**British Library Cataloguing-in-Publication Data**
A catalogue record for this book is available from the British Library.

**THE WONDERS OF SCIENCE**

ISBN 978-981-12-9198-2 (hardcover)
ISBN 978-981-12-9199-9 (ebook for institutions)
ISBN 978-981-12-9200-2 (ebook for individuals)

For any available supplementary material, please visit
https://www.worldscientific.com/worldscibooks/10.1142/13807#t=suppl

Typeset by Stallion Press
Email: enquiries@stallionpress.com

*To Dvora, my beloved wife,*
*Companion and friend throughout my life*

# Contents

# Introduction

The average person is not generally aware of how wondrous the universe is. Everything appears to be quite mundane. The sun rises in the East and sets in the West; the moon illuminates the night sky; water boils at 100 degrees and freezes at zero degrees; there are 24 hours in a day and 365 days in a year; the seasons change over the course of the year; there are towering mountains and vast oceans on our planet. None of this seems particularly wondrous or even impressive. These facts are taken for granted.

What does impress the average person is the tremendous *technological* advancement that has occurred during the last century, which includes man's landing on the Moon, nuclear-powered submarines, the Internet, Google and email, smartphones, and cars that drive themselves. However, the average person would be hard-pressed to name even one important *scientific* advancement that occurred during the last century.

That is what this book is all about. We shall describe some of the wondrous *scientific* discoveries made during the last century, particularly in the field of physics. These amazing discoveries reveal that the simple eighteenth-century orderly universe conceived by Galileo, Laplace, and Newton has been replaced by a universe so complex and so amazing that it almost defies comprehension. We begin by listing some of the remarkable findings of physics that were discovered during the last century, the explanations of which will be presented in the following chapters:

- It was assumed by the great physicist Isaac Newton that clocks everywhere in the universe tell the same time because space and time are independent of each other. However, it was shown by Albert Einstein that space and time are intimately connected and are jointly called *space-time*. As a result, the rate at which time advances depends on the location of the clock. Near a large mass, time will advance more slowly. Thus, a clock on the sun advances more slowly than a clock on Earth.

  Also, the rate at which time advances depends on the motion of the clock. For example, if a clock is moving rapidly past me, the time measured by this clock will advance more slowly than the time measured by a stationary clock that I hold in my hand.

- It had always been a basic principle of physics that if one performs the same experiment twice, under the exact same conditions, one will obtain the same results. However, this basic principle has now been overturned. According to quantum theory, even if an experiment is performed twice under the same exact conditions, one may nevertheless obtain different results each time.

  The reason for this surprising result is that the laws of physics are *not deterministic*. Deterministic laws imply that there is only one possible result for any experiment. According to quantum theory, the laws of physics are *probabilistic*. This means that there are various possibilities of *what might occur* as the result of an experiment, with each possibility having a certain probability of occurring. One can calculate what these various possibilities are and also the probability of the occurrence of each. However, there is no way to know in advance which of the various possibilities *will happen in practice*.

  If one repeats the same experiment, the same set of possibilities apply. But in the second experiment, a different one of the various possibilities may occur.

- It was previously thought that if one knows the initial location of a moving object and the forces that are exerted on that object, and if one then measures its final location, one can use the laws of physics to *deduce* the entire path of the object. It is now known that this is not the case. The only way to determine the path of the object is by *measuring the path directly*.

- Before the twentieth century, physicists knew of only three elementary particles: protons, neutrons, and electrons. However, over the last century, *twenty-six* different elementary particles have been discovered. Also, each elementary particle has a "partner," called an *anti-particle*. The *anti-particle* is identical to the particle, except that its charge has the opposite sign.

- It was long thought that matter is conserved. That is, matter can be transformed from one form into a different form, but the *amount of matter* can never be increased or decreased. Similarly, it had been thought that energy is conserved. That is, energy can be transformed from one form into a different form, but the *amount of energy* can never be increased or decreased. However, Einstein showed that *matter can be created from energy and that energy can be created from matter.*

- Elementary particles were long considered to be basic entities of the universe. A remarkable discovery has shown that the basic entities of the universe are not particles at all but rather *submicroscopic "strings."* These strings vibrate (like a violin string) and the energy of vibration *appears* to us as a particle.

- Personal experience tells us that the universe consists of three spatial dimensions: up–down, right–left, and forward–backward. But physicists have discovered that the universe actually consists of *ten* spatial dimensions, the three familiar dimensions plus *seven additional dimensions*. These additional seven dimensions are so tiny that they can never be observed *directly*. However, their existence can be established *indirectly*.

- It was previously thought that gravity was a force, similar to the electric force. It is now known that gravity is not a force at all but rather *a distortion of space*. This *distortion of space* causes objects to move and gives the impression that a force exists between objects.

- It is now known that the world's weather is *extremely sensitive* to the motion of even a relatively few air molecules. In fact, the motion of a few air molecules *anywhere* on the planet can significantly influence the weather *everywhere* on the planet. The prime example of this is the following: If a butterfly flaps its wings in Singapore, the resulting slight motion of air molecules can eventually influence the weather in New York. This discovery is known as the *butterfly effect.*

- Astronomical measurements of the rotation rate of galaxies have shown that there exists a *previously unknown source of matter* that is completely different from all other forms of matter. Moreover, the amount of this newly discovered source of matter is *much greater* than all the matter previously known. This new source of matter is termed *dark matter.* As of this writing, physicists have no idea regarding the nature of dark matter.

- Measurements of the motion of distant galaxies have demonstrated that there exists a *previously unknown source of energy* that tends to push the galaxies apart. This newly discovered source of energy is called *dark energy.* It acts as *anti-gravity*, repelling galaxies from each other, whereas "regular" gravity causes galaxies to attract each other. As of this writing, the nature of dark energy remains unknown.

- As is well known, electrons repel each other, according to the principle "like charges repel." However, this principle has always been puzzling. Since the electrons have no contact with each other, how can they affect each other's motion? The puzzle was resolved by the discovery that an electron constantly emits tiny "balls" of energy, called *photons.* As an emitted photon collides with a nearby electron, the photon "pushes" the electron away. Thus, the photon repels the nearby electron.

*Each and every one of the above-mentioned scientific discoveries contradicts knowledge of physics that had been universally accepted prior to the twentieth century.* In the coming chapters, we explain the scientific discoveries that have led to these wondrous results.

One must never forget that although the study of science is carried out by great scientists, these scientists are human beings who are subject to human failings. On occasion, blunders have been made by even the greatest scientists, including Isaac Newton, Albert Einstein, and Charles Darwin. We will describe some blunders in the fields of physics and evolutionary biology.

It should be emphasized that these remarkable scientific findings can be explained without the use of mathematics. There will be no mention in this book of Hilbert spaces, Hermitian operators, eigenvalues, or any of the rest of the technical jargon that fills academic books on physics. Nobel laureate Richard Feynman, the great expositor of physics, once remarked that if he could not explain a theory in a way that his mother could understand it, then he himself has not properly understood the theory. (No offence is intended to Mrs. Lucille Feynman!)

The book is organized into several parts. Part 1 deals with physics. Part 2 deals with evolution. Part 3 deals with the blunders made by great scientists.

# Part 1

# The Wonders of Physics

A wonder fills one with awe, upon learning about or seeing something beautiful, astounding, and extraordinary.

# Chapter 1

# The Submicroscopic Universe

In our everyday lives, we observe only the macroscopic aspects of our universe. However, there also exists a microscopic universe that contains items that cannot be seen by the naked eye, but only through a microscope. These microscopic items include bacteria, blood cells, and tiny specks of dust.

But underlying the microscopic universe, there exists an extremely remarkable submicroscopic universe that contains particles that are even smaller. Scientists are aware of these particles through their significant influence on the macroscopic universe. Some submicroscopic particles are familiar, such as electrons, atoms, and protons. However, many submicroscopic particles are unfamiliar to the layman, including quarks, neutrinos, bosons, and deuterons. Although they are extremely small, the submicroscopic particles are crucially important and they form the building blocks of our universe.

This chapter contains an account of our growing knowledge of the wondrous submicroscopic universe. We will see that with each new discovery, an additional layer of complexity of the universe is revealed. Indeed, the submicroscopic universe has proved to be far more complex and far more remarkable than anyone had imagined.

## Particles

The universe consists of two basic components: matter and energy. In this first chapter, we discuss matter, whose elementary building blocks are particles.

By the eighteenth century, scientists realized that the thousands of different types of materials were all composed of different combinations of a few basic particles that they called "atoms."

(The concept of an "atom" was introduced by the Greek philosopher Democritus, but his ideas were very different from the modern scientific concept of an atom.)

The same atom can appear in very different forms and in very different materials. For example, oxygen atoms are a component of the atmosphere (gas), a component of water (liquid), and also a component of rocks (solid). Although the atmosphere, water, and rocks are very different materials, nevertheless, they all contain the same atoms of oxygen, but in different proportions.

Over the years, nearly a hundred different types of atoms, also called chemical elements, have been discovered. This discovery was an enormous advancement in our understanding of nature, reducing the multitude of various materials to only about one hundred different chemical elements.

It was previously thought that the atom was a basic particle that did not consist of smaller particles. (The Greek word "atom" indicates this. The first syllable "*a*" means "not," as in "apolitical" or "asexual," and the second syllable "*tom*" means "to cut." Thus, the word "*atom*" means an entity that "*cannot be cut*" into smaller pieces.)

By the end of the eighteenth century, it was realized that the minute atom consists of even smaller particles. The picture of the atom that emerged resembles a miniature solar system. The central core of the atom — *nucleus* — contains two types of particles, *protons and neutrons*. Around this nucleus, revolve *electrons*. Thus, the atom consists of three types of particles: *protons, neutrons, and electrons*. Protons have a *positive* electric charge, electrons have a *negative* electric charge, and neutrons have *no electric charge*.

Atoms differ from each other according to the different number of protons contained in the nucleus. The nucleus of the hydrogen atom contains one proton, the helium nucleus contains two protons, the lithium nucleus contains three protons, and so on, until the uranium nucleus, which contains 92 protons. Thus, there are 92 different types of atoms, called elements.

The universe of the early twentieth century thus seemed relatively simple, with all the different types of matter consisting of only three different kinds of particles: protons, neutrons, and electrons.

# Radioactivity and the Neutrino

In 1896, the phenomenon of *radioactivity* was discovered by Henri Becquerel and was then developed further by the husband-and-wife team of Marie and Pierre Curie, each of whom was awarded a Nobel Prize. They discovered, to their astonishment, that there are certain atoms that are not stable. The nucleus of these unstable atoms spontaneously emits or "radiates" some of its protons and neutrons. An atom having such an unstable nucleus is termed *radioactive*. Radium and uranium are two well-known examples of radioactive atoms.

Physicists discovered that the phenomenon of radioactivity indicates that there must exist an *additional particle*, which is called the *neutrino*. (The connection between radioactivity and the neutrino is beyond the scope of this book.)

Thus, the universe proved to be even more complex, containing *four* different particles (*electron, proton, neutron,* and *neutrino*). But this is only the beginning of our story.

# Anti-particles

In 1928, Nobel laureate Paul Dirac developed a new theory of physics that combined quantum theory with Einstein's theory of relativity. Dirac's theory demonstrated that every particle has a "partner," called an *anti-particle*. The anti-particle has the *same* properties as the original particle except that it has a charge of the *opposite* sign. For example, the anti-particle of the electron is the anti-electron, which is identical to the electron except that the anti-electron has a positive charge, whereas the electron has a negative charge.

# Quarks

By the middle of the twentieth century, physicists discovered that neither the proton nor the neutron was an elementary particle. (An elementary particle is a particle that cannot be broken down into smaller particles.) The proton and the neutron are each composed of *even smaller particles*, which are the true elementary particles. These even smaller elementary particles are called *quarks*. There are two types of quarks, called the "up" quark and the "down" quark (and, of course, their anti-quarks). Both types of quarks are charged particles. The "up" quark has a positive charge of 2/3, whereas the "down" quark has a negative charge of –1/3.

The proton and the neutron each consist of three quarks. The proton consists of two "up" quarks and one "down" quark, whereas the neutron consists of one "up" quark and two "down" quarks.

(The name *quark* was chosen by Nobel laureate Steven Weinberg, who had proposed the existence of quarks. The word appears in *Finnegan's Wake* by James Joyce. The relevant line in the novel is as follows: *"Three quarks for Mister Mark."* Since the proton and the neutron each consist of *three* of these particles, the whimsical Weinberg thought that *quark* would be an appropriate name.)

The universe now seemed to consist of the *four* following elementary particles: *"up" quark, "down" quark, electron, and neutrino.* But our story is just beginning.

## Surprise!

Just when scientists had thought that they understood the universe, a *fifth* elementary particle was discovered in 1936! Upon hearing of this, Nobel laureate Isidor Isaac Rabi asked his famous question: "Who ordered that?" In other words, "Why does the universe contain a fifth elementary particle?" Every known phenomenon could be explained within the framework of just four elementary particles.

This fifth elementary particle was found to have exactly the same properties as the electron, except that it is about 200 times heavier. It is basically a *heavy electron*.

More surprises soon followed. Scientists discovered there was also *a heavy neutrino, a heavy "up" quark, and a heavy "down" quark*, thus yielding *eight* different elementary particles. Scientists then discovered *a still heavier electron, a still heavier neutrino, a still heavier "up" quark, and a still heavier "down" quark*. This yields *twelve* different elementary particles. (It can be demonstrated that there does not exist another set of even heavier elementary particles). Each group of four elementary particles is called a *family*. Thus, there exist three families of elementary particles.

A tantalizing question remains. Why are there *three* families of elementary particles? Why does nature appear in triplicate? All known phenomena of nature can be explained in terms of only one family of four elementary particles.

To summarize, the universe consists of twelve different elementary particles, arranged into three families: three types of electrons, three types of neutrinos, three types of "up" quarks, and three types of "down" quarks. This picture of the universe seemed to be complete — but it wasn't, as we shall see.

## The Forces Between Particles

Forces act between particles. Two of these forces are familiar to us in our daily lives: the force of *gravity* and the *electromagnetic force*. It was once thought that the magnetic force and the electric force were *two separate forces*. However, in the 1860s, Scottish physicist James Clerk Maxwell demonstrated that the *electric force* and the *magnetic force* were different aspects of the *same force*, now called the *electromagnetic force*.

In addition, there also exists a *third force* in nature, called the *strong nuclear force*. The existence of the strong nuclear force can be demonstrated as follows. The atomic nucleus consists of several protons that are very tightly bound together. Since each proton has a positive electric charge, and positive charges repel each other, why doesn't the electromagnetic force split the nucleus apart? What force binds the protons together so tightly in the nucleus?

The extreme stability of the atomic nucleus is due to an additional force — the *strong nuclear force*. This force causes protons and neutrons to be very strongly attracted to each other. The strong nuclear force that *attracts* the protons to each other is much stronger than the electromagnetic force that tends to *repel* the protons from each other. Therefore, the atomic nucleus is extremely stable. This third force is called a *nuclear* force because it acts only within the tiny nucleus of the atom. By contrast, the force of gravity and the electromagnetic force act even over long distances.

We have so far discussed three forces: *gravity, electromagnetic force, and strong nuclear force*. However, this simple picture proved to be incomplete. The universe had other surprises in store.

## The Weak Nuclear Force

In 1896, the phenomenon of *radioactivity* was discovered. This phenomenon indicates that there exists an *additional force* in nature, which is called the *weak nuclear force*. (The connection between radioactivity and the weak nuclear force is beyond the scope of this book.)

Thus, there are *four* different forces in nature (*gravity, electromagnetic force, strong nuclear force,* and *weak nuclear force*). Once again, the universe proved to be more complex than previously thought.

## Grand Unification Theory

Scientists always try to simplify the description of nature by combining the separate forces, showing that they are simply *different aspects of one single force*. For example, James Clerk Maxwell demonstrated in the 1860s that the electric force and the magnetic forces are really two aspects of one single force, known as the "electromagnetic force." He also showed that the phenomenon of light was another aspect of this same force, with light waves really being "electromagnetic waves." Thus, electricity, magnetism, and light are not *three* different phenomena, but only *one*. This was a great simplification.

Physicists have continued with this program. In the twentieth century, they showed that three of the four forces of nature are just different aspects of one single force. The unification of the electric force, the weak nuclear force, and the strong nuclear force is called "*Grand Unification Theory*." This discovery was an important advancement in our understanding of the universe

# Higgs Particle

Grand Unification Theory predicts the existence of yet another elementary particle, called the *Higgs particle* (after Scottish physicist Peter Higgs, who had proposed it). However, at the time that it was proposed by Higgs, this particle had not been detected.

If the Higgs particle were to be detected, it would serve as striking confirmation of the Grand Unification Theory. In 2012, physicists announced that they had obtained definitive evidence for the existence of the Higgs particle. Thus, the Grand Unification Theory was confirmed. The discovery of the Higgs particle was considered so important that Peter Higgs was awarded a Nobel Prize.

# Boson Particles

Particles attract or repel each other through the transmission between them of special particles, called *boson particles*. (The term *boson* is taken from the name of the Indian Nobel-Prize-winner Satyendra Nath Bose.) For example, electrons repel each other through the electromagnetic force because the electrons transmit *boson particles* to each other. The transmitted boson particles push the electrons apart. The *boson particles* that cause the electromagnetic force are the *photons*. Each separate force is caused by the transmission of specific *boson particles*.

The strong nuclear force is caused by *eight boson particles* called *gluons*, so named because the strong nuclear force "glues" the protons and neutrons together very tightly in the nucleus. The weak nuclear force is caused by *three boson particles* called $W^+$, $W^-$, and $Z^0$. The force of gravity is caused by *one boson particle* called the *graviton*.

# Standard Theory of Elementary Particles

In the course of their studies, scientists discovered 13 different *boson particles*. In addition, there are the 12 elementary particles previously discussed. Finally, there is the Higgs particle.

This complex picture is known as the *standard theory of elementary particles*. Nevertheless, it is widely believed that the *standard theory of elementary particles* cannot be the final word on the subject. Physicists are convinced that there must exist some underlying structure that relates these two dozen elementary particles to each other. It seems inconceivable that so many different types of elementary particles could exist without being related to each other in some way. It is thus not surprising that some of the world's leading physicists are working hard to reveal this underlying structure. This program is known as *going beyond the standard model*.

## Appendix: Producing Elementary Particles

Elementary particles can be produced by means of a large facility called an accelerator. In order to produce the elusive Higgs particles, physicists constructed a giant accelerator in Geneva (known technically as the *large hadron collider*). The main purpose of the giant Geneva accelerator was to produce Higgs particles.

An accelerator produces particles in the following way. An electrically charged particle, say, a proton, is placed within an electric field which accelerates the proton. The proton is constrained to move in a circular orbit (which can be arranged by means of magnets). Each time the circular orbit of the proton brings it back into the electric field, the proton will receive another "push" and thus will go even faster.

The accelerator functions like a person pushing a swing. The swing is pushed forward each time it returns to the "pusher." But unlike a proton in an accelerator, the swing does not go faster with each push because friction slows it down. In an accelerator, the protons move in a vacuum where there is no friction. Therefore, with each "push" by the electric field, the protons move even faster. When many protons are accelerated together, they form a "beam" of speeding protons.

A moving particle has kinetic energy, the energy of motion. The beam of rapidly speeding protons contains an enormous amount of kinetic energy. In the accelerator, two such beams of protons are formed, with the beams carefully aligned to move in exactly opposite directions and they are aimed to collide head-on. For this reason, physicists often refer to an accelerator as a *collider*.

When two speeding cars collide head-on, the collision causes the cars to stop, and thus they lose all their kinetic energy. The supposedly "lost" kinetic energy goes into smashing up the cars. Similarly, in an accelerator, when the two proton beams collide, the protons stop. But unlike the colliding cars, the supposedly "lost" kinetic energy from the proton beams produces new particles.

The production of particles from kinetic energy follows from Einstein's equation, $E = Mc^2$, where $E$ denotes energy, $M$ denotes matter (in the form of particles), and $c$ denotes the speed of light. Einstein's equation states that matter ($M$) can be converted into energy ($E$). This is the basis for nuclear energy, destructive nuclear bombs, and productive nuclear power stations.

Einstein's equation works in both directions. Not only is it possible to convert matter into energy but it is also possible to convert energy into matter. This is what the accelerator does. It converts the enormous "lost" kinetic energy of the speeding proton beams into matter in the form of particles.

The higher the kinetic energy of the speeding proton, the greater will be the mass of the newly formed particle. Producing a very heavy particle requires a very energetic beam. In order to accelerate the protons to extremely high kinetic energy, the accelerator must generate huge electric fields that are perfectly aligned. This is why the Geneva accelerator is so complicated — and so expensive.

The giant Geneva accelerator consists of a vast underground tunnel that is about 20 meters in diameter and 27 kilometers in length. This tunnel is crammed with complicated scientific instruments. The giant Geneva accelerator is the largest and most expensive scientific facility thus far built.

According to the Grand Unification Theory, the Higgs particle is too heavy to have been produced by existing accelerators which is why this particle had not been detected previously. But the Geneva accelerator is indeed powerful enough to produce Higgs particles. In fact, this was one of the primary tasks of the Geneva accelerator. Accordingly, in 2012, it was announced that physicists had obtained definitive evidence for the existence of Higgs particles, which served to confirm the Grand Unification Theory.

It should be pointed out that physicists did not detect the Higgs particle *directly*. This particle is extremely unstable and decays so rapidly that it can never be detected directly. But when the Higgs particle decays, it produces a specific set of decay products, which *are* detectable. Therefore, when physicists detected *exactly* the expected decay products, they rightly concluded that they had detected the Higgs particle.

# Chapter 2

# Dark Matter and Dark Energy

As the twentieth century drew to a close, physicists did not expect any new surprises regarding particles or energy. Although they had not yet detected the Higgs particle, physicists were quite certain that this particle existed and would soon be detected. A similar situation applied to energy. Physicists thought that they had identified all sources of energy. Therefore, it came as quite a surprise when a new type of particle and a new source of energy were discovered.

## Dark Matter: A New Type of Particle

The universe consists of galaxies, which are large clusters of many billions of stars. Galaxies rotate because of the gravitational attraction of their stars toward each other. The rate of rotation of a galaxy depends on the number of stars in the galaxy. However, a careful study showed that the observed rate of the galactic rotation is significantly faster than expected. What could be the cause of this unexpected rapid rotation of the galaxies?

The rapid rotation of galaxies can only be explained if the galaxy contains much more matter than previously thought. In other words, in addition to its stars, each galaxy must contain some sort of additional matter that is not visible to astronomers. This additional matter is termed *dark matter*, with the word *dark* denoting both the unknown *dark* nature of the particles and the fact that these particles do not reflect light as do

13

other particles. This lack of reflection makes the particles *dark* and invisible to the astronomer.

The most astonishing aspect of this dark matter is its sheer quantity. The measured rate of galactic rotation showed that dark matter comprises *80% of all the matter in the universe.* Thus, it turns out that for hundreds of years, physicists have been studying only a minor component of the mass of the universe. They were completely unaware of the major component — *dark matter.* The surprises never cease!

## The Particles of Dark Matter

The particles of dark matter are *different* from other known particles. Physicists seek these particles by means of an instrument called an accelerator, which is capable of producing particles. (The Appendix of the previous chapter explains how an accelerator works.) It is hoped that the powerful Geneva accelerator will produce the particles of dark matter. But as of this writing, the particles of dark matter have not been detected.

## The Age of the Universe

People have long wondered about the age of the universe. According to the universally accepted Big Bang theory of cosmology, the universe had a beginning. When did this beginning occur?

An important advancement was made in 1929 by astronomer Edwin Hubble at the Mount Wilson Observatory near Pasadena, California. Hubble discovered that the universe is expanding. Since the time of the Big Bang, all the galaxies are flying apart from each other at a constant speed.

There is a simple relationship between the speed of an object and its distance from the observer. If an object is moving at a constant speed, and one knows how long the object has been moving, a calculation will tell one how far away the object must be: distance = speed × time. Similarly, if one knows how far away the object is now and also knows its constant speed, then a calculation tells one when the object started its journey: time = distance ÷ speed.

According to the above-mentioned equation, by measuring the speed of a galaxy and how far away it is, one can determine when the galaxies started moving outward, or equivalently, how long ago the Big Bang occurred. This would be the age of the universe.

It is easy to measure the speed at which a galaxy is moving outward. But it is difficult to accurately measure the location of a distant galaxy. However, recent advances in measuring techniques (the details need not concern us here) have provided reliable values of the location of distant galaxies. These data lead to the value of *13.8 billion years for the age of the universe*.

## Dark Energy: A New Source of Energy

As stated earlier, galaxies move outward at a constant speed. The reason is that each galaxy initially received a very big outward "push" due to the Big Bang. After that initial outward "push," nothing else occurred to change the speed of the galaxy. However, this statement is not quite correct. The galaxies attract each other due to the force of gravity, which tends to slow them down somewhat. This effect is very small but measurable. (The effect of gravity is so small because the galaxies are very far apart from each other.)

When this small effect was carefully measured at the end of the twentieth century, physicists were astonished to find that *the speed of distant galaxies is actually increasing slightly*. This discovery was so remarkable that it was awarded the Nobel Prize.

What makes distant galaxies move faster? There seems to exist a repulsive force that pushes the galaxies apart. This *repulsive* force is stronger than the force of gravity that *attracts* the galaxies to each other. Hence, the galaxies are slowly speeding up.

In summary, there appears to be an additional source of energy in the universe that overcomes the attractive force of gravity and pushes the galaxies apart. This additional source of energy is called *dark energy*. There have been many proposals to explain *dark energy*, but as of this writing, no proposal has achieved consensus.

## Surprises

We see that the universe still has surprises in store for us. And these surprises do not relate to minor details. They raise fundamental questions and show that though we have learned much about the universe, we still lack an understanding of some of its very basic features. Studies of the universe will always remain an exciting and challenging topic. One never knows what new surprises await.

# Chapter 3

# Quantum Theory: The Future Does Not Exist

## The Magic of Quantum Theory

Quantum Theory (QT) is the strangest theory in the history of physics. It overturns the following basic principle of science: *If one performs a measurement twice, under the same conditions, the result must be the same in both cases.* However, this principle is incorrect. According to QT, if one performs the exact same experiment twice, the result of the two measurements *may be different.* It follows that the result of a measurement *cannot be predicted in advance.* In other words, *the future does not exist.*

This statement must certainly lead one to raise an eyebrow. Doesn't science assume that every future event can be predicted, at least in principle, from the laws of physics? However, this "obvious" assumption has been overturned by QT.

It should be emphasized that many future events can be predicted with *very near certainty*, say, with 99.999999999...% certainty. Such events are indistinguishable from events that can be predicted with *100% certainty.* This is why QT was not discovered until the twentieth century. But it is now known that many events cannot be predicted even with near certainty. *This is the magic of quantum theory.*

## Classical Physics

Before discussing QT, it is useful to first summarize the laws of classical physics as they were known in the nineteenth century, before the advent of QT.

According to classical physics, the future behavior of physical objects is already fixed in the present. In other words, *the present determines the future*. If one knows all the details of a system at the present time, and one is clever enough to solve the known equations, *one can predict the future behavior of that system*. This picture of the world gave rise to the notion of a "clockwork universe." Given the present position of the hands of a clock, the cogs, wheels, and gears of the clock mechanism determine where the hands of the clock will be at any instant in the future. In the same way, the present situation and the laws of physics completely determine the future of the universe. This was the bedrock of classical physics.

## The Situation in Physics in 1900

The progress of physics had been so rapid and so comprehensive that by the end of the nineteenth century, it was widely believed that all the basic laws of physics were known. A famous example is the statement by Lord Kelvin (William Thomson), one of Britain's leading scientists. In his 1900 address to the British Association for the Advancement of Science, Lord Kelvin declared the following:

> *"There is nothing new to be discovered in physics now. All that remains is to make more precise measurements."*

Kelvin's extreme statement may seem naïve today. However, in 1900, he had good reasons for his bold statement.

## Fundamental Discoveries in Physics Before 1900

The universe consists of objects that range in size from the extremely small (atoms and molecules) to the extremely large (planets and stars).

One of the main tasks of physics is to explain how objects move. Explaining planetary motion was one of the basic problems in physics during the Middle Ages. Finally, in 1609, planetary orbits were accurately described by Kepler's laws. These laws state that each planet revolves around the sun in an elliptical orbit, with each planet sweeping out equal areas at equal times. However, no one could explain *why* the planets move in this fashion.

This problem was solved by Isaac Newton in 1687 in his famous book, *Principia*, the most important book of physics ever written. Newton formulated the fundamental law of motion, $F = Ma$. This law states that if force $F$ is exerted on a particle of mass $M$, the particle will move with acceleration $a$.

According to this law of motion, in order to know how an object will move, one must know the force that is exerted on the object. Newton solved this problem for the planets by formulating the law of gravity, the force that the sun exerts on each planet. By combining his laws of motion with his law of gravity, Newton fully explained planetary motion. Moreover, Newton's laws also explained terrestrial phenomena, such as the tides.

Although great progress in physics had been made, important problems remained. These included explaining the electric and magnetic forces, and the phenomena of light, heat, sound, and thermodynamics.

In the 1860s, James Clerk Maxwell showed that the electric force and the magnetic force are two aspects of the same force, now called the electromagnetic force. Maxwell showed that light is also due to this same electromagnetic force. Thus, Maxwell demonstrated that the electric force, the magnetic force, and light are all different aspects of the same phenomenon. This was a very important advancement.

It was also discovered during the nineteenth century that both sound and heat are due to the motion of air molecules. If the molecular motion is coherent (molecules moving in unison), sound is produced. If the molecular motion is incoherent (molecules moving randomly), heat is produced. The laws of thermodynamics were also established, including the law of the conservation of energy.

Given all this progress, it seemed that Lord Kelvin's optimism was justified. There indeed seemed to be nothing fundamental left to discover in physics.

## Classical Physics: Determinism

According to classical physics, every event can be predicted in advance using the laws of physics. This implies that *the present determines the future*, a concept that is called *determinism*. If one knows the present configuration of a system and the forces exerted on the system, then the laws of physics enable one to calculate the behavior of that system for all times in the future. However, this is only true *in principle*. In practice, *technical difficulties* often prevent one from calculating the future behavior of a system.

## Classical Physics: Problems

For hundreds of years, classical physics was successful in explaining every observed phenomenon. However, near the end of the nineteenth century, several phenomena raised questions that classical physics seemed unable to answer. The following are some examples.

## Radiation

When a piece of metal is heated, it radiates energy. At sufficiently high temperatures, the metal radiates so much energy that it begins to glow. As the temperature continues to increase, the metal first glows red, then blue, and finally white.

In the nineteenth century, physicists learned how to calculate the amount of energy radiated by a piece of metal as a function of the temperature. But the calculated energy did not agree with the measurements. Even worse, according to the calculation, *at every temperature, an infinite amount of energy should be radiated by every piece of metal*! This result is clearly absurd. Something was very wrong with the laws of physics.

## Structure of the Atom

In the early twentieth century, the New Zealand physicist Ernest Rutherford proposed the structure of the atom that is known as the

Rutherford model. According to this model, an atom consists of a small central core, called the *nucleus*, which contains positively charged particles (*protons*), around which revolve negatively charged particles (*electrons*). The atom is thus similar to a miniature solar system. Just as the force of gravity holds the planets in their orbits around the sun, so the electric force holds the negatively charged electrons in their orbits around the positively charged nucleus.

There is a serious problem with the Rutherford model. The electrons that revolve around the nucleus are *charged* particles. According to classical physics, when a charged particle moves in a circle, it radiates energy. If the electrons radiated energy, they would fall into the nucleus and the atom would collapse. But atoms are perfectly stable and they do not collapse. Therefore, something was clearly very wrong with the laws of classical physics.

## Solution to These Problems

To solve these problems, and many others, quantum theory (QT) was formulated. There was a lot of confusion regarding QT in its early days because the theory was so strange. It took about 30 years for QT to become fully developed, from 1900 until about 1930. Over the course of several decades, Nobel Prizes were awarded left and right as the different pieces of this strange theory were discovered.

## Quantum Theory and Probability

Unlike classical physics, QT is *probabilistic*. This means that according to QT, *before performing a measurement, one cannot predict the result of the measurement.* One can calculate which results are possible and one can also calculate *the probability* of obtaining each of various possible results. But it is *impossible to know* which of the various possible results will occur in practice. In other words, *the future is not determined in the present.* This principle completely contradicts *the determinism of classical physics.*

Consider the following example. A QT calculation may show that the possible results of a measurement are F, G, or H. But the laws of QT do not tell us *which of the three possible results (F, G, or H)* will be obtained upon performing the measurement. *In fact, this information does not exist.*

In complete contrast to the above-mentioned scenario, classical physics asserts that there is *only one possible result* for every measurement. According to classical physics, for the measurement under discussion, which of these three results (F or G or H) will be obtained is *already determined before one performs the measurement.*

## The Probability Aspect of Quantum Theory: Radium as an Example

We begin with a bit of background. Thorium, element 90, is radioactive. This means that every atom of thorium will eventually decay to form a different chemical element, which is radium, element 88. Radium is also a radioactive element. Every atom of radium will eventually decay to form the chemical element radon, element 86.

Now we come to the point. How much time elapses before a radium atom decays? *No one knows!* Why is that? One only has to place some radium atoms on a table and observe how much time passes before the radium atoms decay. However, if one carries out this experiment, one obtains a strange result. *The time for each radium atom to decay is different.* One radium atom may decay after five minutes, whereas another radium atom may not decay for a thousand years.

According to classical physics, this result is impossible, for the following reason. Since all radium atoms are identical, they must all behave in the same way. This means that the time for each radium atom to decay must be the same. However, according to QT, in spite of the fact that all radium atoms are identical, the decay time is different for each radium atom. The reason for this strange behavior is that *the future (when the radium atom will decay) is not determined in the present (now).* This is the magic of QT.

## When is Quantum Theory Important?

The reader may be wondering why this dramatic result — *that the future is not determined in the present* — had not been observed earlier. In fact, our everyday experience teaches us *just the opposite*. Throughout our lives, we observe that the future *is indeed* determined by the present. Consider a free throw in basketball, meaning that in certain circumstances, a player is permitted to try to throw the ball into the basket without interference from the opposing players. Every basketball player knows that when he is entitled to a free throw, if he throws the ball accurately in the direction of the basket (the present), in a few seconds (the future), the ball will enter the basket to the applause of the crowd. Why do athletes, as well as the rest of us, remain unaware of QT in our daily lives?

The explanation is that the effects of QT are significant *only* when describing the behavior of *extremely minute* particles, on the scale of atoms and molecules. (In the present context, a speck of dust weighing a trillionth of a gram is a *large* particle.) However, when dealing with macroscopic objects, such as basketballs, the difference between the prediction of quantum theory and that of classical physics is extremely small.

When a basketball is thrown in the correct direction, classical physics predicts that the ball will enter the basket with 100% certainty, whereas QT predicts that the chances of the ball entering the basket are 99.9999999…%, with only an extremely small chance that the ball will miss the basket. Since the difference between these two predictions is immeasurably small, an athlete need not be aware of QT in order to become a basketball star.

In everyday life, we are rarely aware of atoms and molecules. We normally deal with large macroscopic objects. So, it might seem that quantum theory has no effect on our daily lives. But this is not the case.

For example, all modern semiconductor electronics is based on the quantum behavior of individual electrons. The quantum behavior of electrons is responsible for the existence of the transistor. According to the principles of classical physics, transistors could not exist. Thus, if classical physics were correct, there could be no modern electronics, meaning no computers, no television sets, no smartphones, no integrated circuits, no space travel, no jet planes, no lasers, no advanced medical apparatus, and the list goes on and on. Modern electronics is only one example of how quantum theory plays a decisive role in our daily lives.

# Chapter 4

# Quantum Theory: Knowledge Does Not Exist Without a Measurement

## Measurements

One of the basic tasks of physics is to explain how objects move. According to classical physics (physics before 1900), if one knows where a moving object begins its journey, what forces are exerted on the object, and where the object ends its journey, the laws of physics enable one to determine the entire path taken by the object. This fundamental statement of classical physics is rejected by quantum theory (QT). According to QT, even if one knows the initial position and final position of a moving object, *this information does not enable one to know anything about its path. Without a direct measurement of the path, there is no knowledge about the path.*

In this chapter, we will elaborate on this very remarkable QT result and discuss its important consequences.

## The Nature of Light: Waves or Particles?

The phenomenon of light has always mystified people. Indeed, physicists have long debated the nature of light. Does light consist of a series of waves or does light consist of a stream of particles? The essential difference between waves and particles is the following: A wave is an entity that is spread out. A particle is a localized entity.

In the seventeenth century, Isaac Newton maintained that light consists of a stream of particles, whereas his distinguished colleague Christian Huygens maintained that light is a wave phenomenon. At that time, it was not possible to decide between these two possibilities, for the following reason. One cannot distinguish between waves and particles through a measurement unless the active element of the measuring instrument is *smaller* than the wavelength of the wave. But the wavelength of visible light is extremely short, and until the nineteenth century, no measuring instrument existed that had such a small active element. Therefore, the controversy between Newton and Huygens remained unresolved.

By the early 1800s, vastly improved measuring instruments became available. It was then found that light has wave-like properties. Throughout the nineteenth century, additional experimental data confirmed that light is indeed a wave phenomenon. And by the end of the nineteenth century, no physicist doubted that light consists of waves.

## The Radical Proposal of Max Planck

Quantum theory began in 1900 with the radical proposal of Max Planck that light is *not* a wave phenomenon after all. Planck was able to explain the puzzling radiation of heated metals (described in Chapter 3) by assuming that light consists of a stream of particles. These particles are called *quanta* (singular: *quantum*, hence the name of the theory) or *photons* (from the Greek word for "light"). For his explanation of the radiation of heated metals, Max Planck was awarded the Nobel Prize.

Planck's proposal that light consists of a stream of *particles* completely explained the radiation of heated metals. However, his proposal contradicted much experimental data that seemed to demonstrate that light is a *spread-out wave phenomenon*.

It should be mentioned that the particle of light — *photon* — introduced by Planck has no mass. Although almost all particles, such as electrons, protons, and neutrons, do have mass, there is no requirement that a particle *must* have mass.

Another experiment involving light that could not be explained in terms of waves was the photoelectric effect. In 1905, Albert Einstein showed that the photoelectric effect can be completely explained if one adopts Planck's proposal that light consists of a stream of particles. For his explanation, Einstein was also awarded the Nobel Prize.

Arthur Compton later carried out another experiment involving light, known as the Compton Effect. This experiment can only be explained by assuming that light consists of a stream of particles. For his experiment and its explanation, Arthur Compton was also awarded the Nobel Prize.

In the following image, Arthur Compton is in his office at the University of Chicago sitting for a portrait by Chicago artist David Bekker, the father-in-law of the author.

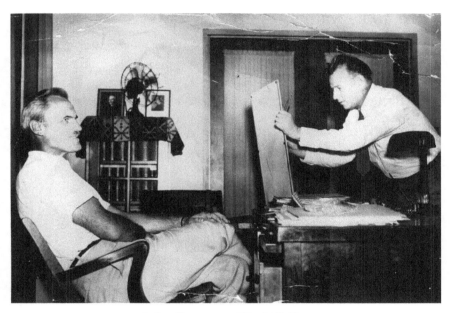

Arthur Compton and David Bekker

## Wave–Particle Duality

We therefore have a very strange and contradictory situation. It appears that some experiments involving light cannot be explained unless one assumes that light consists of *spread-out waves*. But other experiments involving light cannot be explained unless one assumes that light consists of a stream of *localized particles*. Thus, the concept of *wave–particle duality* was born. This contradictory concept asserts that sometimes light behaves like spread-out waves, whereas at other times, light behaves like localized particles.

Such a contradictory situation is impossible. *Light cannot sometimes be localized particles and sometimes be spread-out waves.* Light must *always* be one or the other. According to QT, light *always* behaves like a stream of localized particles and *never* behaves like spread-out waves. As we will see, the concept of *wave–particle duality* is simply a mistake.

## What is the Nature of Electrons?

When the electron was discovered in 1898, it was universally thought that the electron was a localized particle, having measurable particle properties, such as charge and mass. However, in 1924, Louis de Broglie proposed the radical idea that perhaps electrons might sometimes also behave like spread-out waves. He made a comparison between light and electrons. De Broglie pointed out that there are many experiments that seem to show that light consists of spread-out waves. Nevertheless, it was later discovered that, sometimes, light seems to behave like localized particles. Perhaps the same duality also applies to electrons. Although many experiments show that electrons behave like localized particles, there might also be some experiments in which electrons behave like spread-out waves. For this proposal, de Broglie was awarded the Nobel Prize.

When de Broglie's proposal was tested by experiment, it was found that electrons can indeed sometimes *appear* to behave like spread-out waves. The experiments were carried out by Clinton Davisson and George Thomson, who shared the Nobel Prize for their discovery. Thus, the contradictory concept of *wave–particle duality* seemed to apply *both to electrons and to light.*

## The Two-Slit Experiment

The two-slit experiment gave scientists the impression that electrons can sometimes behave as waves. The experimental arrangement for the two-slit experiment is shown in Figures 1 and 2.

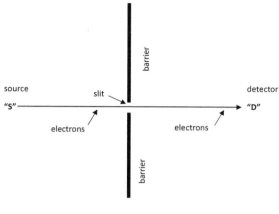

Figure 1

Electrons are ejected, one at a time, from source "S". Each electron moves toward detector "D", which indicates when an electron has reached the detector. Initially, there is a barrier between the source and detector (indicated by the heavy vertical line), which prevents electrons from reaching the detector. A slit is then made in the barrier (indicated by the gap in the heavy vertical line), which enables some electrons to pass through the barrier and then continue on to the detector. In the setup of the experiment (Figure 1), 2% of the electrons emitted by source "S" will pass through this slit and reach detector "D".

Now, we come to the essence of the two-slit experiment. Suppose one now makes a *second slit* in the barrier, as shown in Figure 2.

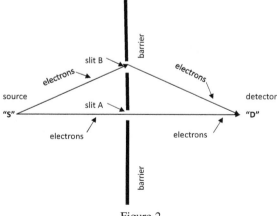

Figure 2

We shall denote the initial slit as slit A and the second slit as slit B. Because of the second slit, in addition to the 2% of the electrons that reach detector "D" by passing through slit A, additional electrons will reach the detector by passing through slit B.

The total percentage of the electrons that reach the detector "D" depends on the distance between slit A and slit B. This is illustrated in Figure 3.

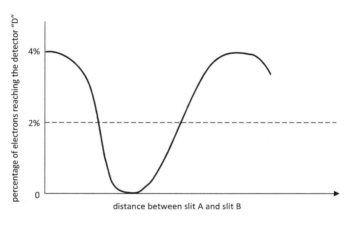

Figure 3

On the horizontal axis is plotted the distance between slit A and slit B. The curve gives the percentage of electrons reaching detector "D" for each given distance between slit A and slit B. Each point on the curve represents a different measurement with the two slits being a different distance from each other.

## Results of the Two-Slit Experiment: Paradox

When the two slits are very close together, one finds that 4% of the electrons reach detector "D", with 2% passing through each slit (see Figure 3). If the two slits are spaced somewhat further apart, the percentage of electrons reaching detector "D" is less than 4%. Since the path from source "S", through slit B, and on to detector "D", is not quite straight, fewer than 2% of the electrons that pass through slit B will reach detector "D".

Now comes the strange result. As the distance between slit A and slit B is increased further, the percentage of electrons that reaches detector

"D" *decreases below 2%!* How can this be? Since 2% of the electrons pass through slit A and reach the detector, how can opening a second slit in the barrier *reduce* the total percentage of electrons reaching detector "D"? Even if *none of the electrons* that pass through slit B reach detector "D", the 2% of the electrons that pass through slit A should still reach the detector. Stranger still, as one sees from Figure 3, for a certain specific distance between slit A and slit B, *no electrons at all reach detector "D"*!

## The Explanation of the Paradox

The original attempt to explain the two-slit paradox was based on the *assumption* that the electron is a spread-out wave that passed through both slits. This explanation is incorrect because it was based on classical physics, which states that one can determine the path of the electron *without making a direct measurement of the electron path*. This assumption is rejected by QT. According to QT, one can only decide whether the electron passed through both slits *by performing a direct measurement of the electron path at the slits*.

One performs the required direct measurement by placing detectors *both* at slit A *and* at slit B. What do these two detectors record?

*The two detectors never record that an electron has passed through both slits*. If the detector at slit A records that an electron has passed through slit A, then the detector at slit B records nothing. If the detector at slit B records that an electron has passed through slit B, then the detector at slit A records nothing. This result shows that the electron passes through *only one slit*. This is the behavior of a *localized particle*.

But what about the paradox? How can it happen that *fewer electrons* reach the detector if *one makes a second slit*? To see how this is possible, one must turn to QT.

## Quantum Theory and the Schrodinger Equation

As discussed in Chapter 3, QT is probabilistic. This means that before performing a measurement, one can never know for certain what the result of the measurement will be. One can only know the *probabilities* of obtaining each of the various possible results.

QT explains how to calculate these probabilities. The basic equation of QT is the Schrodinger equation (SE), whose solution is called the "*probability function*" because it gives the *probability* for an event to occur. In the two-slit experiment, the "event" under discussion is that an electron is emitted from source "S" and reaches detector "D".

The solution to the SE — the probability function — is written as $\psi(x,t)$. Its value depends both on position ($x$) and on time ($t$). However, the value of $\psi(x,t)$ at each $x$ and $t$ is not a number. Rather, it is a *pair of numbers*, commonly called a *complex number*.

The pair of numbers $\psi(x,t)$ can be represented by an arrow lying on a plane (see Figure 4).

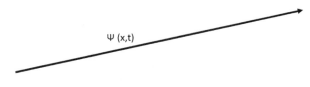

Figure 4

One number of the pair gives the length of the arrow, whereas the second number of the pair gives the direction in which the arrow is pointing.

One must solve the SE for the two-slit experiment. In the present situation, *no forces are exerted on the electron*. The solution of the SE in the absence of forces is the following.

As the particle moves forward, *the length of the arrow does not change*. However, the arrow rapidly changes its direction by spinning around in the plane. The rate of spinning depends on the speed of the electron. For an electron moving at 100 centimeters per second, the arrow spins around about 10,000 times per centimeter of forward motion of the electron. For particles of light (*photons*), the rate of spinning depends on the color of the light. For red light, the spinning rate is about 10,000 times per centimeter of forward motion of the *photon*. For blue light, the spinning rate is about twice as fast.

## Probability of Finding the Particle at a Given Point

The probability of finding the particle at any given point $x$ depends only on *the length of the arrow* at that point. The probability is *unrelated* to the direction of the arrow.

Returning now to the two-slit experiment (Figure 2), the electron can travel from the source "S" to detector "D" via two different paths. The electron can go from source "S" through slit A and then on to detector "D" or the electron can go from source "S" through slit B and then on to detector "D". How does one handle this situation?

This situation is handled in the following way. One calculates the arrows that correspond separately to *each* path, calling them $\psi_A$ and $\psi_B$ (Figure 5).

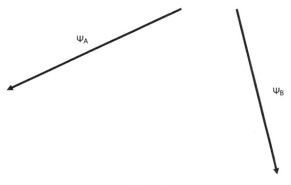

Figure 5

Since the arrow spins extremely rapidly as the electron moves forward, even a slight difference in path length leads to a significant difference in the direction of the arrow. However, both arrows will have the same length, since the length of the arrow does not change as the electron moves forward.

To obtain the total arrow $\psi$(total), one adds the two arrows, $\psi_A$ and $\psi_B$, head to tail, as shown in Figure 6.

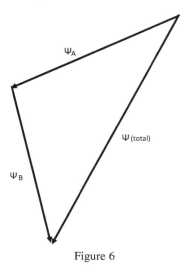

Figure 6

$$\psi(\text{total}) = \psi_A + \psi_{B.}$$

*The length of $\psi$(total) gives the probability that the electron will reach detector "D".*

## Resolving the Paradox

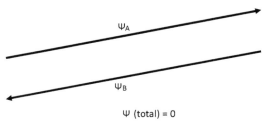

Figure 7

We now consider the situation shown in Figure 7, where the two arrows $\psi_A$ and $\psi_B$ point in *opposite directions*. This will happen if the two slits are a specific distance apart. For this specific situation, the two arrows cancel each other out. That is, adding the arrows $\psi_A$ and $\psi_B$ (see Figure 7) head to tail leads to a total arrow $\psi$(total) of zero length! *This means that if the two slits are a specific distance apart, the electron has zero probability of reaching detector "D".* This is precisely what the measurement shows! Thus, this strange result of the two-slit experiment is explained while assuming that the electron is a localized particle that follows the rules of quantum theory.

## Summary

A lot of ground has been covered in this chapter. Let us summarize the main points:

- A fundamental principle of quantum theory states that knowledge does not exist without a direct measurement. In particular, no valid conclusions can be drawn about the behavior of electrons or light that are not based on a direct measurement.

- Electron behavior follows the laws of quantum theory, which is probabilistic. The probabilities are determined by the Schrodinger equation.
- Both electrons and light *always behave as particles and never as waves*.
- The Schrodinger equation enables one to calculate electron behavior and thereby explain every aspect of the two-slit experiment.
- There is no experiment — neither for electrons nor for light — that requires one to assume a wave phenomenon in order to explain the experiment. The concept of *wave–particle duality* is simply incorrect.

# Chapter 5

# Black Holes

## Introduction

The term *black hole* usually conjures up an image of some dark prison. In the world of physics, however, the term *black hole* refers to something else entirely. A black hole is a concentration of mass so enormous that nothing can escape from it. We will explain why nothing can escape from a black hole. We will also explain why this concentration of mass is called a *hole* and why the associated color is *black*. *Black holes* are yet another example of the wondrous entities that fill our universe.

## Escape Velocity

In order to explain what is meant by the term *black hole*, one must first discuss the concept of *escape velocity*.

If one throws an object upward, the object will soon reach its maximum height, and then, it will fall back down due to the gravitational attraction of the Earth. The greater the initial speed of the object, the greater will be the maximum height of the object before it falls back down.

There is an initial speed so fast that if the object is thrown upward with this initial speed, the object will *escape* the gravitational pull of the Earth and *never fall back down*. This required initial speed is termed the *escape velocity* ("*speed*" and "*velocity*" are synonyms). If an object is thrown upward with a speed that exceeds the *escape velocity*, it "escapes" from the gravitational pull of the Earth and never returns. The more massive the astronomical body, the stronger its gravitational pull, and hence the faster the escape velocity. The following are examples:

- The *escape velocity* from the surface of the Earth is 11 km/s.
- The *escape velocity* from the surface of the sun is 618 km/s.
- The *escape velocity* from the surface of the moon is 2.8 km/s.

An important feature of the escape velocity is that it decreases with height. The higher one is above the surface of the Earth, the lower the Earth's gravitational pull. Hence, it is easier for the object to "escape," implying that the escape velocity is lower. For example, at 1,000 km above the surface of the Earth, the escape velocity is only 10 km/s, whereas the escape velocity is 11 km/s at the surface of the Earth.

## The Relationship Between the Escape Velocity and a Black Hole

It was discovered in 1971 that an extreme concentration of mass exists at the center of our galaxy. Think of a mass ten times the mass of the sun concentrated into the size of Singapore. What would happen if the escape velocity of this extreme concentration of mass exceeds the speed of light (300,000 km/s)? Einstein's special theory of relativity states that *nothing can travel faster than the speed of light*. Therefore, nothing could ever escape from this extreme concentration of mass because no object can ever attain the required escape velocity.

Now, imagine some astronomical body moving in our galaxy. If this astronomical body moves close to the extreme concentration of mass at the center of the galaxy, the astronomical body will be attracted by its gravity. As the astronomical body approaches the concentration of mass, *its escape velocity increases* because the gravitational attraction increases. Finally, when the astronomical body moves sufficiently close to the extreme concentration of mass, its escape velocity will exceed the speed of light.

But according to the special theory of relativity, the astronomical body can then never escape, because nothing can move faster than the speed of light. The astronomical body has become trapped! It is as if the astronomical body has fallen into a *hole. The extreme concentration of mass thus acts like a hole from which nothing can ever escape.*

This is also the case for light. Light always travels at a fixed speed of 300,000 km/s. However, the escape velocity of this extreme concentration of mass may be faster than the speed of light. Therefore, once light enters this concentration of mass, it can never escape. Since no light can escape, this extreme concentration of mass will appear *black*. This is the scientific origin of the term *black hole*.

At one time, there had been a dispute about whether *black holes* actually exist. They seemed to be so fantastic. However, astronomical observations carried out in 1971 convinced scientists that *black holes do, in fact, exist*. It is now widely thought that almost every galaxy has a black hole at its center.

# Chapter 6

# Gravity: The Theories of Newton and Einstein

## Newton's Theory of Gravity

The first theory of gravity was formulated by Isaac Newton in 1687 in his important book, *Principia*. The reader may wonder what people thought before Newton's theory. Didn't everyone know that apples fall from the tree to the ground? Doesn't this imply that there must be some force pulling the apple down?

It is indeed possible to propose an explanation of why objects fall without the need to assume the existence of a force that pulls the objects down. For example, one of the earlier explanations of why objects fall is as follows: It was once thought that there are four elements in nature. In modern terminology, these elements are solids, liquids, gases, and fire. According to this erroneous theory, every object is formed from a combination of these four elements. Each of the four elements was thought to have a strong tendency to move toward its *natural* place. The *natural* place of solids and liquids was thought to be the center of the universe, whereas the *natural* place of gases and fire was thought to be as far away as possible from the center of the universe.

At this time, the center of the universe was thought to be the center of the Earth. According to this erroneous theory, solids and liquids were thought to fall down in order to be as close as possible to their natural place, namely, the center of the universe. Falling down was *not* thought to be caused by a force that is pulling the solid or liquid down.

If one boils water, one sees gas bubbles rising upward through the water. This supports the erroneous idea that these gas bubbles are moving toward their *natural* place, which was thought to be upward, as far away as possible from the center of the universe. The same is true for fire. One observes that flames rise.

Newton's theory was revolutionary. It was a new idea to think that *every object in the universe attracts every other object.* Newton's theory of gravity was able to explain the long-standing puzzle of planetary motion, as well as terrestrial phenomena, such as the tides. (Tides are caused by the gravitational force that the moon exerts on the Earth.)

However, there was a problem. According to Newton's theory, when the gravitational attraction of the sun causes the Earth to move, there is no contact between the sun and the Earth. Therefore, how is the force of gravity transmitted from the sun to the Earth? Newton was well aware of this question. However, he chose to leave this question for future generations to solve.

## Einstein's Theories of Special Relativity and General Relativity

No new theory of gravity was proposed between the time of the theory of Newton (1687) and the time of the theory of Einstein (1915). Therefore, we jump to the twentieth century and turn to Einstein's theory of gravity. There is a very interesting feature of Einstein's theories. He did not formulate his theories in order to explain some experimental data that were not understood. Rather, Einstein formulated his theories in order to answer certain *conceptual questions about nature*, which were unrelated to experiment. This was the case for both Einstein's theory of relativity and his theory of gravity.

Einstein's theory of special relativity is based on the following idea. (We shall soon see the importance of the word *special* in the title of Einstein's theory.) When two objects move relative to each other in a straight line and at constant speed, there is no way to determine which object is moving and which object is stationary. Picture a train moving through a station. The person on the station platform views himself as standing still and views the train as moving forward. However, a person on the train might equally well view himself as standing still and view the station platform as moving backward. There is no way to distinguish between these two points of view.

Einstein asserted that *all laws of physics* must correspond to this important idea. Einstein was convinced that when two objects are moving relative to each other in a straight line and at a constant speed, *it must be impossible, on the basis of the laws of physics, to determine which object is moving and which object is stationary*. This is a *conceptual idea* unrelated to any experimental data.

Maxwell's law of electromagnetism, formulated in the 1860s, was not consistent with this idea. For this reason, in 1905, Einstein formulated a new theory of space and time with which Maxwell's law of electromagnetism would be consistent. This new theory was Einstein's *theory of special relativity*.

## Some Remarkable Results That Follow from Einstein's Theory

Einstein's theory of special relativity led to the following remarkable results:

- Newton had assumed that time was universal. That is, a clock advances at the same rate everywhere in the universe and this rate does not depend on the motion of the clock in space. However, according to Einstein's theory of special relativity, *the rate at which a clock advances depends on the motion of the clock*. If a clock moves past me, the moving clock will advance slower than a stationary clock that I am holding in my hand.

  It should be emphasized that at slow speeds, this effect is extremely small. (A speed of *1,000 km/s* is considered *slow* in this context.) Therefore, it had not been previously noticed that a moving clock advances more slowly than a stationary clock. Only when the speed of the clock approaches the speed of light, namely, about *300,000 km/s*, does this effect become significant.

- Physicists had previously thought that mass was conserved. This means that the total amount of mass in the universe can neither be increased nor decreased.

The same was thought to be true for energy. It had previously been thought that the total amount of energy in the universe can neither be increased nor decreased.

Einstein showed that the *sum* of the total amount of mass *plus* the total amount of energy in the universe is the quantity that is conserved. This sum can never be changed, but the amount of each separate component of the sum (mass or energy) *can be changed.*

- Another result of Einstein's theory is that there exists a maximum speed at which objects can move. This maximum speed is the speed of light (300,000 km/s).

This result seems quite strange. Why can't one simply exert a force on a moving object and thus make it move faster? Newton's laws state that if a force is exerted on a moving object, its speed will increase. According to Newton, if one continues to exert a force, the speed of the object will continue to increase *without limit.*

According to Einstein, however, this is not correct. As an object moves faster, a *greater force is required to increase its speed.* And as the speed of the object approaches the speed of light, an *infinite force* is required to further increase its speed. But an infinite force does not exist. This limits the maximum speed at which an object can move.

## What is *Special* about the "Special Theory of Relativity"?

Why did Einstein refer to his theory as the *special* theory of relativity? This theory is limited to the *special* case in which objects are moving in *a straight line and at a constant speed relative to each other.*

Einstein wondered what laws of motion would apply when an object *is not moving at a constant speed or it is not moving in a straight line.* For example, as a planet travels around the sun, the planet is not moving in a straight line nor is it moving at a constant speed. What laws apply to such motion?

Once again, Einstein was dealing with a *conceptual problem.* At that time, there were no data regarding planetary motion that required a new explanatory theory.

It took Einstein ten years, from 1905 until 1915, to solve this difficult problem. He referred to his new theory as the *general* theory of relativity.

The theory is *general* in the sense that it applies to *all* types of motion, including accelerated motion and curved motion. The important Russian physicist Lev Landau has described Einstein's theory of general relativity as *"the most beautiful theory in all of physics."*

## Einstein's Theory of Gravity

Einstein applied his theory of general relativity to formulate a new theory of gravity.

*Einstein's general theory of relativity is essentially a new theory of gravity.*

The most amazing result of Einstein's theory of gravity is that *gravity is not a force*. Since one is used to thinking of gravity as a force, one must make a switch in one's thinking to understand Einstein's theory of gravity. According to Einstein, gravity is a *distortion of space*. The gravitational attraction between two objects is *not* due to one object pulling on the other object, as is the case for the electric force. Instead, the effect of gravity is that the first object *distorts* the space around it and the second object *moves* in reaction to this distortion of space. Since the *distortion of space* is not visible, *it appears to us as though* the two objects attract each other by means of a force.

**DISTORTION OF SPACE**

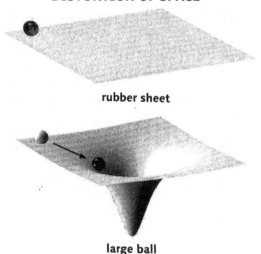

**rubber sheet**

**large ball**

The concept of a distortion of space can best be illustrated by means of the accompanying figure. The top part of the figure shows a stretched rubber sheet on which a small ball lies motionless. The bottom part shows the situation after a large ball has been placed at the center of the rubber sheet. The effect of the large ball is to distort the rubber sheet, with the distortion being greatest in the vicinity of the large ball.

As a result of the distortion of the rubber sheet, the small ball moves toward the point of maximum distortion (which is where the large ball lies), as shown in the figure. Thus, the small ball moves *toward* the large ball, in spite of the fact that *there is no force of attraction between the two balls*. The motion of the small ball is caused by the *distortion* of the rubber sheet.

The *rubber sheet* represents space whose *distortion is invisible to us*. We see only the small ball moving toward the large ball. Because of this, Newton thought that a force of attraction (gravity) exists between the two balls. But Einstein showed that the correct description of gravity is a *distortion of space* (as illustrated by the *rubber sheet and its distortion*), rather than a force.

Einstein's theory of gravity answers the question that was left unanswered by Newton's theory: How is the force of gravity transmitted from the sun to the Earth?

According to Einstein's theory of gravity, the *distortion of space* extends outward from the sun and reaches the Earth. This is the *direct contact* between these two astronomical bodies. If the sun were to disappear suddenly, it would take several minutes until the *distortion of space* disappeared at the position of the Earth. The Earth would *still feel* the effects of the sun's gravity for several minutes after the sun had disappeared.

Einstein's theory of gravity also predicts a *distortion of time*. Amazingly, *time advances more slowly* in the neighborhood of a massive object. A clock located on the sun will advance more slowly than a clock located here on Earth. This slowing down of time (called *dilation of time*) is an extremely small effect. Even on the surface of the very massive sun, *a clock advances only one minute slower over the course of an entire year* than a clock located on the surface of the Earth.

The theories of gravity proposed by Einstein and Newton are fundamentally different. In practice, however, the *numerical predictions* of these two theories are very similar. In fact, the predictions of these two theories are so similar that for two centuries, no one doubted Newton's theory. Whenever the two theories differ in their predictions, it is always Einstein's theory that agrees with the observations. Today, Einstein's theory of gravity is accepted by all physicists.

To summarize, gravity is not a force. Gravity is a *distortion of space and time*.

## Important New Predictions of Einstein's Theory of Gravity

Einstein's theory of gravity makes a number of important predictions that differ from Newton's theory. Two of these predictions were very important historically:

- According to Newton's theory, gravity has no effect on light. Therefore, starlight will always travel in straight lines even when passing near the sun. But according to Einstein's theory of gravity, light rays emitted from stars should bend slightly when passing near a massive body, such as the sun.
- Newton's theory of gravity predicts that the orbit of every planet is an ellipse. But according to Einstein's theory of gravity, the orbit of a planet should differ very slightly from an ellipse. This very small effect is measurable only for Mercury, the planet nearest the sun.

## Observation Confirms Einstein's Theory: Bending of Light Rays

Einstein published his theory in 1915, in the middle of the First World War. At that time, people did not have physics on their minds. But when the war ended in 1918, physicists began to wonder whether Einstein's theory agreed with observations.

As stated previously, according to Einstein's theory, rays of starlight bend very slightly when passing near a massive object. However, the effect is large enough to be measured only when starlight passes very close to the sun. This situation occurs in the daytime. But in the daytime, starlight is masked by the glaring brightness of the sun. As we know, starlight is visible only at night. But at night, the sun is so far from starlight that the bending of starlight is too small to be observed.

However, there is one situation in which starlight passes right near the sun but it is not masked by the sun's brilliance. This situation occurs during a total eclipse of the sun. A total solar eclipse was predicted for 29 May 1919 on the island of Principe, off the west coast of Africa. It was decided that Einstein's theory would be tested during this eclipse.

Very accurate instruments for measuring the position of stars were brought to Principe. Measurements were carried out during the total solar eclipse. If the starlight would bend, the position of the star would appear to be slightly displaced from its *official* position in the sky. The measurements showed that *starlight that just grazed the sun was indeed bent by exactly the amount predicted by Einstein's theory of gravity.* Einstein instantly became a global sensation! He achieved rock star fame, a position that he has retained ever since.

## Observation Confirms Einstein's Theory: Orbit of the Planet Mercury

According to Newton's theory of gravity, the orbit of each planet is an ellipse that repeats itself forever. Elliptical planetary orbits were first reported by Johannes Kepler in 1609. However, no one could explain *why* the shape of the planetary orbits was an ellipse until Newton's theory of gravity provided the explanation in 1687.

According to Einstein's theory of gravity, the elliptical planetary orbit will *not* repeat itself over and over again. Rather, the ellipse should rotate very, very slowly, due to the effect of the massive sun. All planets except Mercury are too far away from the sun for this very slight rotation of the ellipse to be measured.

Only for Mercury, the planet closest to the sun, is the effect large enough to be observed. Even for Mercury, the rotation rate of the elliptical orbit predicted by Einstein is extremely small, only *43 seconds of arc in a century.*

It is worth emphasizing how truly small this effect is. A circle is divided into 360 degrees, a degree is divided into 60 minutes, and a minute is divided into 60 seconds. Thus, the predicted rotation of the ellipse of Mercury is *only 0.005% of a complete circle in 100 years.* It is amazing that such a small effect could be measured at all!

I had previously mentioned that Einstein did not develop his theories to explain data. But here we seem to have a case in which an explanation was needed for the observed data, namely, the very slow rotation of the ellipse of the orbit of Mercury.

The measured rotation of the ellipse of the orbit of Mercury is *570 seconds of arc per century*, and not *43 seconds of arc in a century.* Newton's prediction that the ellipse never rotates at all applies *only* to the sun and one planet. But aside from Mercury, there are seven other planets in the solar system whose gravitation affects Mercury. There is also interplanetary dust that affects the orbit of Mercury. Including all these effects, the prediction of Newton's theory was that the ellipse of the orbit of Mercury should rotate for *530 seconds of arc per century.*

Thus, there is only a 7% difference between the data (570 seconds of arc per century) and Newton's prediction (530 seconds of arc per century). No one had considered this small difference to be a problem. Perhaps there was somewhat more interplanetary dust than had been assumed.

No one thought that the modest 7% difference between the data and Newton's theory indicated a shortcoming in Newton's theory of gravity. But the 7% discrepancy in the rotation rate of Mercury's ellipse was explained by Einstein's new theory. *Einstein had solved a problem that no one, including Einstein himself, had known to exist!*

## Gravitational Waves: Another Confirmation of Einstein's Theory of Gravity

Electromagnetic waves, also known as light waves, are produced by rapidly moving an *electric charge* back and forth. These waves, which can be produced in a TV studio, travel through space and are received by the TV set in one's living room.

According to Einstein's theory of gravity, the same effect should occur because of gravity. If a *very large mass* moves back and forth rapidly, *gravitational waves* should be generated and these waves will travel through space. However, gravity is an extremely weak effect. Therefore, the generated gravitational waves will also be very weak and extremely difficult to detect. A greatly improved measuring apparatus was developed, and in 2015, physicists succeeded in detecting the extremely weak gravitational waves.

# Chapter 7

# Quantum Gravity, String Theory, and the 10-Dimensional Universe

The twentieth century witnessed the development of quantum theory. Quantum theory (QT) has been confirmed in innumerable important experiments and is now a basic principle of physics. Every theory of physics *must* be compatible with QT.

Richard Feynman was awarded a Nobel Prize for developing a procedure, known as "renormalization," that makes a theory of force (the electromagnetic force and the two nuclear forces) compatible with QT. However, Feynman's procedure cannot be applied to Einstein's theory of gravity because, according to Einstein, gravity is not a force, but rather a distortion of space (see Chapter 6). It is this feature that makes Einstein's theory of gravity incompatible with QT.

A compatible theory, called *quantum gravity*, requires a new framework for describing the universe.

## String Theory

The new scientific framework that provides a theory of quantum gravity is known as *string theory*. (Although the most modern version of this theory is known as *M*-theory, we will continue to use the more popular name of "string theory".) String theory is a revolutionary conceptual framework for describing the physical universe.

According to the previous framework, the basic entities of the universe are particles — electrons, quarks, photons, etc. String theory makes the astounding assertion that the basic entities of the universe are not particles at all, but rather tiny "*strings*". These strings vibrate (like a violin string) and the energy of vibration *appears* to us as a particle having mass. Einstein's equation $E = Mc^2$ relates energy to mass.

String theory can be formulated for various numbers of dimensions. The familiar three dimensions in our universe are right–left, forward–backward, and up–down. In three dimensions, however, string theory does not yield a theory of quantum gravity. In fact, for any number of dimensions of space *fewer than 10*, string theory fails to yield a consistent theory of quantum gravity. *But for a 10-dimensional universe, string theory does yield a consistent theory of quantum gravity.*

In summary, physicists discovered that *Einstein's theory of gravity is compatible with quantum theory only (i) in the framework of string theory and only (ii) if the universe consists of 10 dimensions of space.* Therefore, physicists have come to the amazing conclusion that *the universe must consist of 10 spatial dimensions.*

But how can one reconcile the 10-dimensional universe predicted by string theory with our everyday experience of a three-dimensional universe? What meaning can be given to the extra seven dimensions?

## Compacted Dimensions

The three familiar dimensions of space (right–left, forward–backward, and up–down) are infinite in extent. That is, one can move up forever or to the right forever. There is no limit to these three dimensions. However, the additional seven dimensions are not infinite in extent. In fact, these additional seven dimensions are *so extremely short* that they will *never ever* be accessible to our senses or to our measuring instruments. For this reason, it was previously thought that we inhabit a universe of only three dimensions.

The very short seven dimensions are said to be *compacted*, meaning that they are *very tiny in extent*. According to string theory, the extent of a compacted dimension is a billionth of a billionth of a billionth of the radius of an atom (called the *Planck length*, in honor of Nobel laureate Max Planck). Such a dimension is so short that it can never be detected with any measuring device, not now and not in the future. This is the meaning of a *compacted dimension* — a dimension that exists but is far too small to ever be detected.

**COMPACTED DIMENSIONS**

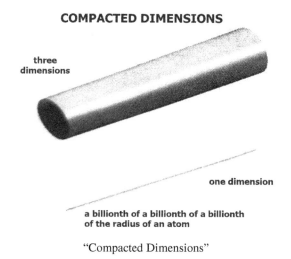

"Compacted Dimensions"

The concept of compacted dimensions is illustrated in the accompanying figure. The top part shows a rod, whose three dimensions are its *length* and its *two-dimensional cross-sectional area*. In the bottom part, the two-dimensional cross-sectional area of the rod has been so diminished that the rod has become reduced to a thin wire. As the two-dimensional cross-sectional area is diminished even further, the two dimensions of the thin wire become observable only with difficulty, and eventually, they cannot be observed at all. But these two dimensions certainly do exist.

Even though compacted dimensions cannot be detected *directly*, they have very significant *indirect* effects on the universe. *The most important indirect effect is that the seven compacted dimensions make possible a theory of quantum gravity.* Explaining other important indirect effects lies far beyond the scope of this book. Because of its many successes, string theory has become widely accepted in recent years.

# Chapter 8

# Chaos Theory: Predicting the Weather and the Butterfly Effect

## Weather Forecasting

Long-range weather forecasting has always been a challenge. Astronomers can predict the exact location of each planet next March 26 at 6:42 pm. But no one can predict what the weather will be in Singapore next March 26 at 6:42 pm. This lack of predictive power for weather conditions is not a technical problem. *It will forever remain impossible to make accurate long-range weather forecasts.* Explaining this surprising statement is the subject of this chapter.

## The Atmosphere

There is no *conceptual* difficulty in long-range weather forecasting. All the components of the calculation are known, as shown in the following:

- The *atmosphere* consists of a *known mixture of gases* (oxygen, nitrogen, argon, carbon dioxide, and water vapor), which, under certain circumstances, form rain.
- The *energy input to the atmosphere* is known. It comes from the sun and the surface of the Earth.
- The *source of atmospheric water vapor* is the oceans and the seas.
- Meteorologists have long understood *the equations that govern the interactions between the various components of the atmosphere.*

55

Since everything about the atmosphere is known, why is there such difficulty in the accurate long-range prediction of rain?

## The Weather — A Non-Local Phenomenon

The enormous practical difficulty in determining future weather stems from the fact that *weather is not a local phenomenon*. The weather in any particular location is strongly influenced by the atmospheric conditions over a surprisingly large area, extending over *many thousands of square kilometers*. For example, Chicago's weather in future days is determined by today's atmospheric conditions *throughout half the United States*. Thus, to predict the future weather of Chicago, one has to calculate seven atmospheric parameters *at thousands of different locations*.

It is obvious that performing so many calculations by hand is quite impossible. Therefore, in the days of hand calculations, long-range weather forecasting did not exist. The prediction of rain was limited to one day in the future.

Since the 1950s, all this has changed. Our TV screens routinely display weather maps containing detailed forecasts extending over several days in advance. The cause of the radical improvement in weather forecasting can be summarized in one word: *computers*. Modern computers can readily calculate millions of numbers.

## A Numerical Example

The difficulty in weather forecasting can be illustrated by means of a numerical example. The following is a very simplified discussion of the complicated science of meteorology.

As stated earlier, in order to predict the weather in Chicago, one must solve the atmospheric equations at every point over half the United States. In practice, one divides the area under consideration into a grid of points placed at intervals of 100 km and 10 layers deep. This corresponds to about 7,000 points and, at each point, one must calculate seven atmospheric parameters. Thus, 50,000 numbers must be calculated.

The principle of prediction is to use the current data to calculate future data. The current data (50,000 parameters) are inserted into equations which are solved by a computer to obtain the atmospheric data at some future time (50,000 new parameters).

Predicting the weather for the day after tomorrow at any particular location requires calculating 10 million numbers! Calculating a single number requires 500 computer operations. Therefore, calculating the weather only two days in advance requires *5 billion computer operations.*

# Computers

We now have an idea of the vast number of calculations necessary for predicting the weather. Various tricks, shortcuts, and climate models have been developed that greatly shorten the computations, but at the end of the day, the computer must still perform billions of operations in order to calculate future weather.

This leads to the fundamental question of whether computers are up to the job. Can today's computers perform the many billions of operations required for long-range weather forecasting?

It should also be mentioned that in addition to great speed, a powerful computer must also have an extremely large memory in order to store all the numbers calculated at each intermediate step of the calculations. However, this aspect of the computer will be ignored here, and we shall concentrate on computer speed. The speed of a computer is measured in terms of its ability to perform a certain number of *operations per second.*

The advancement in computer speed has been nothing short of phenomenal. The first electronic computer (ENIAC) was developed in 1945. By 1965, computers were capable of performing a million operations per second. In 1980, the supercomputer was developed. By the late 1980s, billion-operations-per-second computers became a reality.

Past experience suggests a 10-fold increase in computing speed every five years. Today's engineers are already designing and building the next generation of computers, capable of a *trillion* operations per second.

# Predicting the Weather

Because of the great advances in recent years in computing speed, computers are now capable of performing the many billions of calculations necessary for predicting the weather. This makes it possible for a TV news program to present an accurate five-day weather forecast. As the speed of the computer increases, the predictive power of the meteorologist will increase correspondingly. The fastest computers available today enable scientists to make an accurate prediction of when and where it will rain *for a week in advance.* Armed forces routinely employ such calculations of future weather to prepare military operations.

## Anticipating the Future

As speedier computers become available, even longer-range weather forecasting can be anticipated. The day will surely come when it will be possible to accurately predict the weather for a month in advance, and ultimately, for an entire season in advance. It may require fifty years or maybe even a hundred years, but it seems certain that the day will come when meteorologists will be able to prepare a chart for any specific locality, listing *exactly* which days will be rainy and which will be clear *throughout the coming season.*

For example, the long-range weather chart might proclaim a wet winter for New York, with 9 days of rain in October, 14 in November, 8 in December, 18 in January, 11 in February, and 6 in March. The chart would also list exactly *which* days will be rainy, in exact analogy to a sunrise chart that lists the time of sunrise on each day.

But surprisingly, recent scientific research into weather forecasting has shown that such *long-range weather predictions will never be possible.*

This is not the personal opinion of a few pessimistic scientists. *One can demonstrate that at no time in the future* will it be possible to predict the weather for more than two weeks in advance. Moreover, this statement remains valid even if future computers are developed that are capable of performing a *trillion, trillion operations per second*!

## Chaos

The reason for the inability to predict the future weather can be expressed in a single word: *chaos.* Chaos is a new branch of physics, considered by many researchers to be one of the most important scientific discoveries in recent years.

The concept of chaos should *not* be understood to mean confusion. Quite the contrary. Chaos has its own rules, which lead to well-understood consequences. Chaos occurs in complex systems whose motion is described by what mathematicians call *nonlinear dynamics*. Although the theory of chaos is very complicated, the central idea is readily explained. A chaotic system is a system that is *extremely* sensitive to even the *slightest* changes in the local conditions. By contrast, a non-chaotic system is a system that is not very sensitive to changes in the local conditions. These definitions are best clarified by means of the following simple examples.

## Non-Chaotic Systems and Chaotic Systems

Consider the following non-chaotic system. If one throws a ball, it will land someplace. If one throws the ball again, but this time aiming in a slightly different direction, the ball will land in a slightly different place. Slight changes in the direction of the throw lead to only slight changes in the final landing place. This characterizes a non-chaotic system.

Now consider a system that is chaotic. Inflate a balloon with air and release the air. As the air rushes out, the balloon lurches erratically back and forth in a way that is *impossible to predict*. If one repeats this experiment, *the balloon will never repeat the same jerky motion*, no matter how carefully one tries to duplicate the conditions. The reason lies in the *extreme sensitivity* of the balloon's motion to the initial conditions.

The balloon can never be filled with *exactly the same* amount of air as before to a trillionth of a gram or be pointed in *exactly the same* direction as before to a trillionth of a degree. There will always be *extremely slight differences* between the two experiments, and these *minute* differences are sufficient to cause the second jerky motion of the balloon to be *completely* different from the first jerky motion. The almost unbelievable sensitivity to initial conditions is the hallmark of a chaotic system.

# The Butterfly Effect

The extreme sensitivity to conditions that characterizes a chaotic system has been given a picturesque name. In the case of the weather, meteorologists refer to this phenomenon as the *butterfly effect*. The term graphically expresses the *extreme sensitivity* of the weather to even the *slightest changes* in atmospheric conditions *anywhere* in the world. As amazing as it seems, if a single butterfly flutters its wings in Singapore, this trivial event can have a *major effect* on the weather in New York, on the other side of the planet.

It is obviously impossible to include in the database for a weather forecast the fluttering of the wings of every butterfly everywhere on the planet. It is similarly impossible to include in the database all other minute events that influence the weather.

*Therefore, it will always be impossible to make a long-range weather prediction.*

The butterfly effect does *not* imply that extreme meteorological events will occur. For example, a butterfly in Singapore *will not* cause snow to fall in Tel Aviv in August, because it *never snows* in Tel Aviv in the summer. However, on January 15th, Tel Aviv could experience *either* a sunny, warm day *or* a cold, rainy day. Singapore's butterflies *do* play a role in determining which of these two possibilities will occur.

It should be emphasized that the term *butterfly effect* is not merely a figure of speech. It is *literally true* that the fluttering of the wings of a *single* butterfly will eventually cause large-scale changes in the weather *everywhere* on our planet. This illustrates the extreme sensitivity of the weather to the details of atmospheric conditions.

Our discussion of the butterfly effect shows that developing faster computers will not solve the problem of long-range weather forecasting. The fastest computer of the future will never be able to input *all the data* that influence the weather, including even the slightest movement of every living creature on the planet. Thus, long-range weather forecasting will not be possible even with a computer capable of a trillion, trillion, trillion operations per second!

# Characteristic Time for Chaos

There is a feature of chaotic systems that is extremely important for our discussion. Let us return to the balloon experiment described earlier. The motion of the second balloon will, in fact, be almost identical to the motion of the first balloon for a *very short* time. However, after about *half a second*, the motion of the second balloon will become totally different from the motion of the first balloon.

This teaches us the following: The extreme sensitivity of a chaotic system takes effect *only after* a certain characteristic time — about half a second in the case of the balloon experiment. Therefore, *it is possible* to predict the motion of the balloon *during the first half-second after its release, before this characteristic time*. However, after the first half-second, it will no longer be possible to predict the motion of the balloon. The existence of such a characteristic time is a central feature of chaotic systems.

# The Atmosphere — A Chaotic System

It is now known that the atmosphere is a system that exhibits chaotic behavior. Therefore, there exists a characteristic time after which accurate long-range weather forecasting is impossible.

The key question is the duration of the characteristic time. For some systems, the characteristic time is very short, only half a second in the inflated balloon experiment.

Meteorologists have discovered that the characteristic time for atmospheric chaos is about 12 days. Hence, accurate weather prediction is possible for 12 days in advance.

But longer-range weather prediction is impossible, and will forever remain impossible.

# Chapter 9

# The Solar System: Geocentric *versus* Heliocentric

## The Structure of the Solar System

One of the oldest debates in the history of astronomy relates to the structure of the solar system. Is the sun situated at the center of the solar system, with the Earth and the other planets rotating around a stationary sun (heliocentric system: *helios* is Latin for sun)? Or does the Earth lie at the center of the solar system, with the sun, moon, and the other planets rotating around a stationary Earth (geocentric system: *geo* is Latin for Earth)? Nowadays, it is known that the solar system is heliocentric. But throughout the Middle Ages, the matter was a subject of vigorous debate. It certainly *appears* to anyone standing on the Earth that the sun and the other planets are rotating around our planet.

Important scientific evidence to decide this matter is provided by the phenomenon of *parallax*.

## Parallax

Parallax means the observed change in the apparent position of an object *due to a change in the observer's location.* Consider the following example. One looks at a tree that is a kilometer away and finds with his compass that the tree lies due north of him. If he moves a hundred meters to the east, his compass will now show that the tree now lies somewhat northwest of him. This change in the direction relative to the tree is called *parallax*.

What happens if one were to move only one meter to the east? His compass will still show that the tree lies somewhat to the northwest of him, but much closer to due north than in the previous case when he had moved a hundred meters to the east. If the person were to move only one centimeter to the east, which is only 0.001% of the distance to the tree, then it would require an extremely accurate compass to show that the tree does not lie directly due north.

## Parallax in Astronomy

An astronomer may observe that a star appears in a particular position in the sky. If the astronomer looks at the star again but from a significantly different location, the apparent position of the star will have shifted somewhat. This shift in position is called *parallax*.

How can the astronomer move to a significantly different location? Even the two most distant locations on Earth are only about 12,000 km apart. But this distance is negligibly small compared to the distance to even the nearest stars. In the example of the tree, this distance corresponds to the person moving only one centimeter to the east. Therefore, there will be no measurable parallax.

It turns out that *it is possible* for an astronomer to move so far away that his new location will be significantly different from his previous location. Actually, he doesn't have to move at all! If the heliocentric theory of the solar system is correct, then the astronomer automatically moves from one side of the sun to the other side of the sun every 6 months, as the Earth upon which the astronomer stands revolves around the sun. This corresponds to a new location that is about *300 million km* distant from the previous location. This certainly seems to be a large enough change in the astronomer's location for parallax to be observed for the stars. That is, the apparent position of the stars should shift somewhat throughout the year. But if the geocentric system is correct, then the astronomer hardly moves at all as the seasons change. Thus, there should not be any parallax, that is, no change in the apparent position of the stars.

We have here *a clear scientific test* to enable one to choose between the geocentric system and the heliocentric system. If the stars exhibit parallax as the seasons change, then the heliocentric system is correct. However, if even the nearest stars show no sign of parallax, then the geocentric system is correct.

What did the measurements actually show? *They showed no sign of parallax*! The apparent position of the stars did not change as the seasons changed. This seems to confirm that *the Earth is not revolving around the sun*. The lack of parallax appeared to be strong scientific evidence in support of the geocentric theory of the solar system.

## Refutation

The error in the above-mentioned reasoning is as follows. Until two centuries ago, scientists were unaware of how extremely distant the stars are. It is now known that even the nearest stars, the *Centauri* stars, are over *40 million, million kilometers from the Earth*, an almost incomprehensible distance! Therefore, moving 300 million km as the seasons change corresponds to moving only 0.001% of the distance to the nearest star. In the example of the tree, this corresponds to moving only one centimeter when the tree is one kilometer away. In the Middle Ages, instruments for measuring the position of stars were not accurate enough to detect such small parallax.

It wasn't until 1838, with the advent of more accurate measuring instruments, that mathematician-astronomer Friedrich Bessel succeeded in measuring the tiny parallax of the nearby star *61 Cygni*. Since then, the parallax of many other stars has been measured. Thus, the heliocentric solar system was confirmed.

There is an important lesson to be learned from this history of the parallax of stars. It is often claimed that "one nasty fact" can overturn a theory. However, the "one nasty fact" often turns out to have been a misinterpreted "fact."

# Chapter 10

# Mathematics in the Service of Science

## Mathematics — The Language of Science

The laws of science are always expressed in mathematical terms. Every physics textbook is filled with mathematical equations. For this reason, mathematics is often called "the language of science."

The overriding role of mathematics in physics has long puzzled many scientists. In 1960, Nobel laureate in physics Eugene Wigner published a famous article on this topic, entitled "The Unreasonable Effectiveness of Mathematics in the Natural Sciences." Wigner begins with a parable (somewhat altered here) about two friends who met after a long separation. One friend had become a physicist, and he showed his non-scientist friend an article containing the equations for electricity. When the non-scientist asked about the meaning of a particular symbol, the physicist explained that the symbol ($\pi$) is called Pi and it represents the ratio of the circumference of a circle to its diameter. The non-scientist replied, "You must be joking! What possible connection can there be between the laws of electricity and the geometric properties of circles?"

Wigner went on to write that "one may smile at the simplicity of this question. Nevertheless, I had to admit to an eerie feeling because the reaction of the non-scientist showed common sense." Wigner asked himself why mathematics appears in every expression of the laws of physics.

The following is another example. The basic equation of quantum theory is the Schrodinger equation, which determines the behavior of electrons, atoms, and molecules. The Schrodinger equation contains the symbol $\pi$. Therefore, following Wigner, one may ask, "What possible connection can there be between the equations of quantum theory and the geometric properties of circles?"

If a physicist manages to discover a new law of nature, he will *always use mathematics to express his new law*. Only mathematics has the ability to express complex ideas in terms of a few symbols and equations, whose meaning is readily understood by the community of physicists.

Finally, it should be emphasized that the very existence of laws of nature is by no means obvious. Einstein once remarked, *"The most incomprehensible feature of the universe is that it is comprehensible."*

## What is Modern Mathematics?

The mathematical knowledge of most non-mathematicians is limited to what is taught in high school. The high school curriculum includes courses in algebra, geometry, trigonometry, and often even calculus. The student learns to perform calculations and to solve equations. However, this is *not* what mathematicians study in their *research*.

Modern mathematics consists of esoteric topics, including topology, differential geometry, linear algebra (no connection to high school algebra), group theory, graph theory, set theory, number theory, Calabi-Yau shapes, $M$-theory, and other topics that are completely unfamiliar to the layman. One might think that while these esoteric topics are, no doubt, very interesting to the professional mathematician, they probably have no relevance to the physicist who studies the "real world" of rocks, stars, and atoms. This idea is completely incorrect. It turns out that even the most esoteric mathematical topics often play a vitally important role in the description of various aspects of the "real world" as listed in the following:

- Differential geometry is essential for Einstein's theory of gravity.
- Four-dimensional complex Minkowski space is the most useful framework for analyzing Einstein's special theory of relativity.
- Group theory is essential for understanding the fundamental forces of nature.

- Topology is a very useful framework for formulating the properties of important optical systems.
- Number theory was essential for producing the public-key encryption system that guarantees safe online credit card transactions.
- Game theory has important applications to economics.
- The Kakeya problem in higher dimensions is a very useful framework for describing hydrodynamics.

## The Four-Color Problem: How "Pure" Mathematics Influences the "Real World"

There is a well-known problem in mathematics known as the "four-color problem." Suppose that one wants to color a map such that no two countries that share a common border will have the same color. It is easy to show that three colors are insufficient for the map and that one can color any map with five colors. Moreover, every mathematician was completely convinced that one can *also* color any map with only four colors. But no one was able to prove this. The "four-color problem" became one of the famous unsolved problems in mathematics.

The "four-color problem" is an example of what is called "pure mathematics." Pure mathematics is widely considered to be "useless" and not only by laymen.

What possible importance could result in providing formal proof for something that everyone is already convinced to be true? Moreover, why is it important to know just how many different colors are required to prepare a map?

It turns out that the search for proof of the four-color problem led to important advances in a branch of mathematics known as "graph theory." Graph theory has a very close connection with computer science. In fact, one really cannot be considered a theoretical computer scientist unless one is proficient in graph theory. Thus, the search for a solution to the four-color problem indirectly led to advances in computer science.

The mathematicians who worked so diligently to try to solve the "four-color problem" had no interest at all in computer science. They were working on this problem because of their love of mathematics. Nevertheless, their findings later proved to be of enormous practical value.

To many of us, it may seem that mathematicians are just playing games with their symbols, and in fact, this may well be the case. Nevertheless, important practical results always seem to result from their "games." Therefore, mathematicians should be encouraged and provided with adequate funding, even if most of us have no idea what they are really doing or why they are doing it.

# Chapter 11

# The Fine-Tuned Universe: Why Does the Sun Shine?

## Introduction

Since the 1970s, many scientists have noted that the universe appears to be *fine-tuned*. This chapter is devoted to presenting the evidence that supports this statement. However, we must first clarify what is meant by the term "fine-tuned."

The term "fine-tuned" means fine-tuned for some purpose. With regard to a piano, fine-tuning means that the steel-wire strings have been carefully adjusted so that the piano will produce beautiful music. When applied to the universe, the concept of fine-tuning relates to the fact that the universe embodies the very strict conditions that allow *life*. It is known that "any old universe" could not produce life. Very specific conditions had to be fulfilled in order to produce living creatures, and all these conditions are indeed satisfied in the universe. In this sense, the universe is fine-tuned. Because of its relevance to human life, the fine-tuning of the universe is also called the *anthropic principle*, from the Greek word *anthropos*, meaning "man."

The special conditions necessary for life can be divided into the following three categories:

- If the laws of nature were only *slightly different*, then *life would be impossible*.

71

- If planet Earth were only a *little bit different*, then *life would be impossible*.
- Unique conditions were necessary in order to cause inanimate material to be transformed into living cells.

We shall discuss each of these three topics in turn.

# Solar Energy

A quite remarkable connection exists between the laws of nature and the existence of life. A relationship between the principles of biology and the existence of life is not surprising, but one would not expect to find such a relationship when it comes to physics. However, scientists have discovered that the existence of life is intimately connected with certain specific details of the laws of physics.

It is clear that life on Earth depends crucially on the sun. Without the energy and warmth from the sun, our planet would be incapable of supporting life. Therefore, our discussion of fine-tuning begins by examining the mechanism that produces the sun's energy. (Since the sun is an ordinary star, what will be stated here regarding the sun also applies to every other star.)

The sun contains only two kinds of atoms: hydrogen and helium. Helium is inert and does not undergo any reactions. Therefore, helium need not concern us further. Our discussion centers on hydrogen, the simplest atom, whose nucleus consists of one proton, a fundamental particle of nature.

The sun is basically a vast assemblage of protons. The manner in which protons produce solar energy was explained in the 1930s by Hans Bethe, who was awarded the Nobel Prize for his explanation. (Bethe was a German Jew, and like all German Jews, he was dismissed from his university position by the Nazis. Bethe came to the United States and joined the faculty of Cornell University, where he made his Nobel Prize-winning discovery.)

Bethe's explanation consists of the following steps:

- Because of the extreme conditions in the interior of the sun, a proton occasionally transforms spontaneously into a neutron (another fundamental particle of nature).

- The resulting neutron combines with another proton to form a composite particle called a deuteron.
- Deuterons undergo a complicated thermonuclear reaction, called "thermonuclear burning," which provides the enormous energy and brilliant light of the sun.

Thus, deuterons constitute the "solar fuel" whose thermonuclear burning generates the energy of the sun that enables life to exist on Earth.

An important feature of solar thermonuclear burning is that *it occurs very gradually* because protons rarely transform into neutrons. This ensures that the sun "burns" slowly, generating solar energy very gradually. The sun has been shining for 5 billion years and will continue to shine for another 5 billion years.

Another nuclear reaction that might, only in principle, occur in the sun is the combination of one proton and another proton. Fortunately for us, a proton–proton combination does not occur. If one proton were able to combine with another proton, then all the protons in the sun would rapidly combine with each other. This would produce a gigantic explosion of the sun, which would destroy all the planets in the solar system.

## Fine-Tuning

Both the *possibility* of a proton–neutron combination and the *impossibility* of a proton–proton combination depend on the strength of the nuclear force between protons. Detailed analysis of the nuclear force has led to the following results:

- If the nuclear force were only a *few percent weaker*, then a proton could *not* spontaneously transform into a neutron. As a result, there would be *no* neutrons in the Sun and hence *no* deuterons. Without nuclear fuel, the sun would *not* shine. The sun would be a cold ball of protons, precluding the possibility of life.
- If the nuclear force were only a *few percent stronger*, then the protons in the sun would *rapidly* combine with each other, generating enormous amounts of energy. If this were to happen, the entire sun would soon *explode*, precluding the possibility of life.

Fortunately, *the strength of the nuclear force happens to lie precisely within the narrow range in which neither of these two catastrophes occurs.*

- The proton–proton explosion of the sun *does not* occur.
- The gradual "thermonuclear burning" of deuterons in the sun *does* occur.

This provides the energy, heat, and light that are vital for life on Earth. If the strength of the nuclear force were *just slightly different*, there could be no life on Earth. This is our first example of fine-tuning.

## Water and the Oxygen-Rich Atmosphere on Our Planet

There are two types of planets in the solar system. The nearby planets (Venus, Earth, and Mars) are called "rocky planets" because of their hard, rock-like surface. The more distant, much larger planets (Jupiter, Saturn, Uranus, and Neptune) are called "gas giants" because of their gaseous composition. We shall here be discussing the rocky planets.

It is clear that both water and an oxygen-rich atmosphere are necessary for the existence of life. The Earth is blessed with an abundant supply of both, permitting life to flourish here. Our two neighboring planets, Venus and Mars, are devoid of both liquid water and an oxygen-rich atmosphere. Hence, these planets are devoid of life. These facts may not seem particularly noteworthy, but we shall see how remarkable they really are.

Planetary scientists have discovered that soon after the formation of the Earth, Venus, and Mars, all three planets had large amounts of surface water. The deep channels that are observed today on the surface of Mars were carved out long ago by fast-flowing Martian surface water. Similarly, Venus was once covered by oceans that contained the equivalent of a kilometer-deep layer of water over its entire surface. Eventually, however, all the water on Mars froze and all the water on Venus evaporated.

How did our planet Earth escape these catastrophes that struck Venus and Mars? Our planet avoided these catastrophes through the occurrence of the "greenhouse effect."

Radiation from the sun warms the surface of our planet. The surface of the Earth reflects this heat upward in the form of infrared radiation. Gases in the atmosphere, primarily carbon dioxide, trap the infrared radiation and prevent it from escaping into space. This is called the "greenhouse effect" and the gas that causes this effect is called a "greenhouse gas." The infrared radiation that is trapped causes the surface of the Earth to be much warmer than would be the case without the greenhouse effect.

The atmosphere of Venus contains vast quantities of greenhouse gas, which caused the planet to become so hot that all surface water boiled away. Venus became much too hot to permit life to exist. On the other hand, Mars does not have enough greenhouse gas in its thin atmosphere to keep the surface warm. As a result, the surface of Mars is extremely cold. All the water that was once on Mars has frozen.

The Earth has *just the right amount of greenhouse gas to keep the surface warm, but not so much greenhouse gas that would make the surface very hot.* Therefore, among the rocky planets of the solar system, Earth alone has surface water, permitting life to exist!

The atmosphere of the Earth is controlled by a very delicate balance, involving the subtle interplay of numerous factors. This delicate balance caused planet Earth to develop an oxygen-rich atmosphere. This delicate balance did not exist on either Venus or Mars. Hence, neither of these two planets has an oxygen-rich atmosphere.

This remarkably fortunate history of our life-sustaining planet Earth has been given a special name by scientists: *the Goldilocks problem of climatology.* Recall the children's story in which Goldilocks entered the home of the three bears and found the various items of Baby Bear to be "not too hot and not too cold ... not too hard and not too soft ... not too high and not too low ... but *just right.*" In the same vein, the existence of water and an oxygen-rich atmosphere on Earth is caused by conditions that are *just right* for the existence of life. This remarkable situation is an example of fine-tuning.

## Physics and Astronomy

The above-mentioned examples of fine-tuning are just two of the many instances that could be indicated from the physical sciences. Many scientists have commented on the severe requirements for the existence of life. Particularly perceptive are the words of Freeman Dyson, a physicist at the Institute for Advanced Study in Princeton (where Albert Einstein was a professor for many years):

> *"As we look out into the universe and identify the many peculiarities of physics and astronomy that have worked together for our benefit, it almost seems as if the universe must in some sense have known that human beings were coming."*

## The Origin of Life

Although the laws of physics and astronomy are fine-tuned for the existence of life, that does not *guarantee* that life would actually appear. *Scientific American* has described the extreme improbability of the transformation of inanimate material into living cells.

- Harold Klein, chairman of the U.S. National Academy of Sciences committee that reviewed origin-of-life research, was quoted as follow:

> *"The simplest bacterium is so damn complicated that it is almost impossible to imagine how it happened."*

- Nobel laureate Francis Crick was also quoted:

> *"The origin of life appears to be almost a miracle, so many are the conditions which would have had to be satisfied to get life going."*

If this Nobel laureate chose the words *"almost a miracle"* to describe the conditions necessary for the origin of life, then it is clear that an incredible series of highly unlikely conditions must have been fulfilled to permit an inanimate material to become transformed into life. Our universe is indeed fine-tuned.

# Part 2

# The Wonders of Evolution

# Chapter 12

# Evolution: Denial

Biological evolution has an unusual feature. Poll after poll shows that, in the United States, nearly half the population denies the validity of evolution. The very fact that such polls are conducted tells us that something strange is happening. No one takes a poll to record the views of the average layman regarding the validity of chemistry, physics, or statistical mechanics. Polltakers know in advance that no one will deny the validity of these sciences. But when it comes to evolution, many people think that this science has no validity. Why does evolution generate such opposition?

There are several reasons for the widespread skepticism regarding evolution, such as the following:

- Many religious people are of the opinion that evolution contradicts the Bible. Therefore, they believe that evolution must be wrong.
- Many people, non-religious as well as religious, find the idea that "Man descended from a monkey" very offensive.
- The basic idea of evolution seems quite incredible. How could the one-celled amoebae evolve into a huge elephant or a flying bird? How can the small harmless birds of today be the descendants of the gigantic terrifying dinosaurs of the past?
- The principles of evolution can be expressed in simple terms, without the need for mathematical equations. This leads many people to think that they understand these principles and that they are qualified to express an opinion regarding their validity.

## Fact *versus* Theory

The term *evolution* relates to the *fact* that the vast panorama of animal species that we observe today evolved from animal species that lived in past eras and are now extinct.

*The theory of evolution* proposes a mechanism to explain how ancient species transformed into the modern species that we observe today.

In discussions about evolution, the failure to distinguish between fact and theory is the most common error made by laymen and, unfortunately, not only by laymen.

The most famous theory ever proposed to explain the *fact* of evolution is *Darwin's theory of evolution*. In recent years, competing theories have been proposed, and we shall discuss some of these competing *non-Darwinian theories of evolution*. The average layman will probably be surprised to learn that there exist non-Darwinian theories of evolution.

The term "Darwin's theory" is commonly used to refer to the *combination* of Darwin's original theory *and* the genetic theory of Gregor Mendel, the father of genetics. This combination is also called *neo-Darwinism* or the *synthetic theory* because it is a synthesis of Darwin's original ideas with genetic theory. Charles Darwin was unaware of genetics when he published his theory in 1859.

## The Importance of Darwin's Theory

Darwin's theory was an enormous step forward in our understanding of evolution. His vital contribution must not be minimized. But nearly two centuries have passed since Darwin published his book, *The Origin of Species*, and our understanding of evolution has made great strides forward as more scientific evidence has accumulated. Therefore, one should not be surprised to learn that Darwin's two-centuries-old theory is in need of some revision. The required revision will be discussed in subsequent chapters.

# Chapter 13

# Evolution: Religion

An unfortunate aspect of Darwin's theory of evolution is that some scientists, even respected scientists, think that there are religious implications to Darwin's theory. In particular, it is sometimes claimed that Darwin's theory supports atheism. Consider the following example.

## Darwin's Theory and Richard Dawkins

According to evolutionary biologist Richard Dawkins, of the University of Oxford,[1] *"Darwin made it possible to be an intellectually fulfilled atheist."* There are some very problematic aspects of this assertion:

- No one can imagine that a respected physicist would write that "Maxwell's formulation of electromagnetic theory made it possible to be an intellectually fulfilled atheist." And before Maxwell's theory, was it not possible to be "an intellectually fulfilled atheist"?
- The assertion of Dawkins that Darwin's theory supports atheism is not correct. Others have drawn the *exact opposite conclusion* from Darwin's theory. Consider the words of Rabbi Samson Raphael Hirsch, an important leader of Orthodox Judaism (translated from German and shortened)[2]:

  *"If the concept of evolution were to gain complete acceptance by the scientific world, Judaism would call upon its adherents to give even greater reverence to God, Who in His boundless creative wisdom,*

82 *The Wonders of Science*

*needed to bring into existence only one amorphous nucleus and one law of adaptation and heredity in order to bring forth the infinite variety of species that we know today."*

These words of Rabbi Hirsch clearly demonstrate that Darwin's theory has not transformed Rabbi Hirsch into "an intellectually fulfilled atheist." Far from it!

Particularly interesting is Rabbi Hirsch's assertion that the evolution of the animal kingdom is *even more impressive* than producing every species by a separate act of divine creation. For example, it is impressive to make a beautiful pair of shoes, but it is much more impressive to make a factory that can take raw materials and from them produce (see Darwin's words in the following) *"endless forms"* of shoes *"most beautiful and most wonderful."*

Darwin's theory can be interpreted *either* as supporting atheism *or* as supporting religious belief, depending on who is doing the interpreting.

It is interesting to note what Darwin himself thought about this question. The following lines are from Darwin's final paragraph in *The Origin of Species*:

*"There is a grandeur in this view of life, having been originally breathed by the Creator into a few forms or into one, and from so simple a beginning, endless forms most beautiful and most wonderful have been and are being evolved."*

Note Darwin's word "Creator" with a capital "C." These stirring sentiments of Darwin are definitely not the words of "an intellectually fulfilled atheist."

# References

1. R. Dawkins, 1986, *The Blind Watchmaker* (W. W. Norton: London and New York), p. 6.
2. S. R. Hirsch, 1850–1888, *Collected Writings*, translation published in 1997 (Feldheim Publishers: New York), vol. 7, p. 264.

# Chapter 14

# Challenges to Evolution

## Introduction

There are many deeply religious people who do not see any contradiction between their belief in a Supreme Being and evolution. They accept evolution, just as they accept physics and chemistry, as being the workings of nature, as ordained by God. These religious people view the laws of nature as an expression of God's dominion over the world. In particular, they refer to the scientific principles regarding evolution as "theistic evolution." This term means that evolution occurred, just as stated by evolutionary biologists, but that evolution was directed by God. We have previously quoted one early proponent of this view in the Jewish tradition, namely, Rabbi Samson Raphael Hirsch (see Chapter 13). There are many other proponents of this view in the Catholic tradition, in the Protestant tradition, in the tradition of Islam, and in other traditions.

However, there are also many religious people who think that evolution is incompatible with their religious beliefs, claiming that evolution contradicts the Bible. Therefore, these people do not accept evolution and they look for ways to disprove evolution. They claim that science has shown that evolution is impossible. We shall examine two of these challenges against evolution.

One challenge is known as the "*argument from design*" and the other is known as "*intelligent design*." Despite the similarity of the titles, these two challenges are very different. Their only common feature is that *they are both incorrect.* In this chapter, we shall present these two challenges and explain why each is incorrect.

## The Argument from Design

The challenge of the "*argument from design*" that evolution is impossible dates back 1,000 years. (The word "argument" does not denote a dispute. "Argument" is an old English word for "proof.") A convenient formulation of this challenge is known as the "*watchmaker argument*," proposed in 1802 by the English theologian William Paley[1]:

> "*If one were to find a rock in the forest, one could imagine that the rock had not been made by anyone. However, if one were to find a watch in the forest, no one would suggest that the watch had not been made by a person. The precision with which the cogs, springs, and wheels of the watch have been fashioned, and assembled to serve a particular purpose, demonstrate that the watch could not have been formed by natural processes. Its complexity and specific design prove that the watch must have been made by a watchmaker.*"
>
> "*If one now considers the vast panorama of animals, each consisting of many complex organs that function together in intricate ways to permit each animal to live, one sees far more complexity than is found in any watch. Therefore, if the complex design of a watch requires a watchmaker, so must the complex design of the natural world require a Maker, who must be God.*"

Paley contrasts a "rock in the forest" with a "watch." He states that one can learn nothing about a "Maker" from the rock because rocks result from the laws of nature. Therefore, there is no need to assume the existence of a "rock maker."

Unlike a rock, a watch is a very complex object, consisting of "*cogs, springs, and wheels*," and such complex objects *do not arise* spontaneously from the laws of nature. *The laws of nature never produce a watch.* Since watches do exist and they did not result from the laws of nature, Paley concludes that watches must have been produced by a "watchmaker." Living creatures are even more complex than a watch. Therefore, concludes Paley, living creatures must have had a "maker," who can only be God.

This is the essence of Paley's "watchmaker argument." We will now explain what is incorrect about Paley's argument.

## Complex Objects Form Spontaneously from the Laws of Nature

Paley's *watchmaker argument* is based on the assumption that complex objects never form spontaneously from the laws of nature. However, this assumption is incorrect. Examples follow.

Snowflakes are extremely complex crystals of snow in the form of beautifully intricate structures, with each snowflake being different and each having a perfect six-fold pattern of fractal symmetry. Nevertheless, there is no "snowflake maker." Snowflakes form spontaneously through the laws of nature under certain weather conditions.

There are many other examples of complex objects that form spontaneously through the laws of nature. The chemist can list unbelievably complex molecules that form whenever the required raw materials are present. The physicist can list the complex vertex structure of type-II superconductors. The crystallographer can list the intricate structure of crystals. To summarize, *the complexity of an object does not constitute evidence that it was designed by a conscious being.* In particular, complex living creatures could have been produced spontaneously through evolution.

## Experience, Not Complexity

Another critique of the *argument from design* is the following. If one walks in the forest and finds the letters ABC carved on a tree, one would certainly conclude, and correctly so, that someone had carved these letters. There is nothing complex about the shapes of these few letters — merely a few lines arranged in a simple pattern. But it is clear that some literate person had carved these letters on the tree.

The reason why one knows that letters did not form spontaneously through the laws of nature is that our *experience* tells us the letters ABC *never* form on a tree spontaneously through the laws of nature. Letters on a tree are *invariably* the work of a literate person who carved them.

Not complexity but *experience* tells us whether an object was fashioned by a conscious agent or was formed without any such intervention. One concludes that snowflakes and rocks are not the result of any person's design because one's *experience* indicates that snowflakes and rocks form spontaneously from the laws of nature. Similarly, the reason one concludes that a watch found in the forest was made by a watchmaker is that one's *experience* indicates that watches *never* form spontaneously from the laws of nature. The complex design of the watch is irrelevant to this conclusion.

What about the animal kingdom? Was it fashioned by a conscious being or did it come into existence spontaneously through the process of evolution? One has *no experience* to guide one because there exists *only one animal kingdom*. Therefore, the argument from design does not permit one to conclude *anything* about the origin of the animal kingdom, regardless of its extreme complexity.

This refutes the argument from design. However, it is important to repeat the fact that the animal kingdom arose through the process of evolution does not have any religious implications. In particular, *evolution does not contradict the Bible and does not constitute a problem for a person of faith.*

## Intelligent Design

The *intelligent design* (ID) challenge against evolution is quite recent. ID is the subject of the 1996 book written by biochemist Michael Behe, entitled *Darwin's Black Box: The Biochemical Challenge to Evolution.* Behe, a religious Catholic, claimed to have discovered an *ironclad proof* for the existence of a supernatural being, whom he called the "Intelligent Designer." His studies of the living cell led Behe to conclude that gradual Darwinian evolution cannot explain many of the complex biochemical reactions that take place in the cell. Only ID can explain them. Although Behe refrained from identifying the Intelligent Designer, the widespread understanding is that the Intelligent Designer is God.

Behe's proposed proof that the living cell could not have formed through Darwinian evolution has generated enormous interest (reported in *Newsweek, U.S. News & World Report, New York Times, Commentary, National Review* and other periodicals).

Unlike previous individuals who rejected evolution, Michael Behe is a Professor of Biochemistry at a respected university, a research scientist who performs experiments, has been awarded grants and publishes papers in international science journals. Behe's book is the most sophisticated critique of evolution to appear in recent years.

Soon after Behe published his book, H. Allen Orr, an evolutionary biologist at the University of Rochester, published a definitive proof that *Behe had erred in his claim.* Orr demonstrated that Darwinian evolution *is indeed able* to explain the many complex biochemical reactions that take place in the living cell.

Behe's supposed proof for the existence of a supernatural entity is based on a single claim and *that claim was demonstrated by Orr to be erroneous.*

## *Darwinian Evolution*

Darwinian evolution operates through the appearance of chance mutations in the genetic makeup of an animal. A "favorable mutation" is a mutation that enhances the animal's chances for survival by making the animal a bit stronger, faster or less susceptible to disease, etc. An animal with a favorable mutation has a greater chance than other individuals to live long enough to produce offspring. Therefore, a favorable mutation is likely to become incorporated into the species gene pool. The accumulation in the gene pool of many favorable mutations over many generations brings about extensive changes in the animal, eventually leading to an entirely new species.

The key point of Darwin's theory is that *only favorable mutations that enhance the animal's chances for survival* will become incorporated into the gene pool.

Behe asserts that the gradual accumulation of favorable mutations *cannot* explain the development of many vital biochemical mechanisms.

Among the examples cited by Behe is the mechanism for blood clotting, which is vital for the animal's survival: *Twelve different biochemical reactions are involved concurrently in blood clotting, and if even one of these twelve biochemical reactions does not occur, blood will not clot.*

Behe claims that the complicated mechanism for blood clotting could not have possibly evolved *gradually through a series of mutations, with each mutation providing an additional survival advantage to the animal.* Each such mutation would, by itself, be *useless*. The *complete sequence* of the 12-step blood-clotting mechanism had to appear in the gene pool to be useful to the creature. But the probability is negligible that all 12 required specific mutations would appear in the gene pool *simultaneously*.

Behe's claim for the validity of ID is based on the *single claim* that the *complete sequence* of the 12-step blood-clotting mechanism had to appear in the gene pool *simultaneously*. If his claim is incorrect, then the entire basis for ID is refuted.

## *Refutation*

Shortly after the publication of Behe's book in 1996, evolutionary biologist H. Allen Orr, of Rochester University, published an article[2] that refuted the above claim of ID. Orr demonstrated that the 12 specific mutations needed for blood clotting could indeed have appeared in the gene pool *gradually, one after the other, through intermediate stages, with each intermediate stage providing an additional survival advantage to the animal.* This process is, of course, in complete accord with Darwin's theory.

We present the details of Orr's refutation of Behe's claim in the Appendix.

## The Situation Today Regarding ID

Since H. Allen Orr has conclusively proven that the central claim of intelligent design is incorrect, one would expect that interest in ID would have died down. But a Google search shows that this is not the case. To this very day, there are published articles claiming that ID is correct, with their authors seemingly unaware that the basis for Behe's claim for the validity of ID has been shown to be incorrect. Similarly, some published articles also denounce ID, even though their authors do not appear to be aware of exactly what Behe has claimed. The sterile arguments continue.

## Appendix

Here, we present Orr's explanation of how the 12 mutations necessary for blood clotting could have *entered the gene pool gradually, through intermediate stages, with each intermediate stage providing an additional survival advantage.* This is the case even though *all 12 mutations are required* for the 12-step biochemical reaction that causes the blood to clot.

The following table lists the influence of each mutation on the biochemical system. The term "irreducible system" means that all the mutations must be present in the gene pool for the biochemical system to function.

| Mutation number | Biochemical system | Improvement? | Irreducible system? |
|---|---|---|---|
| 0 | **A** | — | — |
| 1 | **A + B** | Yes | No |
| 2 | **A\* + B** | Yes | Yes |
| 3 | **A\* + B + C** | Yes | No |
| 4 | **A\* + B\* + C** | Yes | Yes |
| — | **—** | — | — |
| 22 | **A\* + B\* + C\* + D\* + E\* + F\* + G\* + H\* + I\* + J\* + K\* + L** | Yes | Yes |

In the distant past, a biochemical system may have consisted of only one part, say, part **A**. The system worked, but not too well. A genetic mutation then produces part **B**, which leads to a somewhat improved system, consisting of **A + B**. This improved system is *not* irreducibly complex because the improved system will function even without part **B**. A second genetic mutation then transforms **A** into **A***, which leads to a further improvement of the system. However — *and this is the crucial point* — **A*** will not work unless **B** is present. Therefore, the system consisting of **A*** + **B** *is* irreducibly complex because *both* **A*** *and* **B** *are necessary for the system to function.*

We have thus shown how an irreducibly complex system can be produced by gradual evolution, with each mutation leading to an improvement in the system, *even though the final system* (**A*** + **B**) *will not function* unless both its parts are present. Therefore, we are done. The claim of ID — *that this is impossible* — has been refuted.

But let us continue. A third genetic mutation now occurs to produce part **C**, which leads to further improvement. This system is *not* irreducibly complex because it will function without part **C**. A fourth mutation transforms **B** into **B***, yielding yet another improvement. However, **B*** will not work unless **C** is also present. Therefore, the system (**A*** + **B*** + **C**) *is* irreducibly complex because *all three parts are necessary for the system to function.* Nevertheless, this irreducibly complex system was produced *by a series of gradual improvements* in accordance with gradual Darwinian evolution.

This process can be continued to produce the following 12-part irreducibly complex system: **A*** + **B*** + **C*** + **D*** + **E*** + **F*** + **G*** + **H*** + **I*** + **J*** + **K*** + **L***.

An important feature of this procedure is its *irreversibility*. After the biochemical system is complete, one cannot determine the order in which its 12 parts were formed or what the intermediate parts were (**A, B, C, D, E, F, G, H, I, J, K, L**). Once the scaffolding has been removed, there is no way to determine how the irreducibly complex building was constructed.

# References

1.   W. Paley, 1802, *Natural Theology* (Fauler: London), vol. 5, pp. 1–2.
2.   H.A. Orr, December 1996, *Boston Review*, pp. 34–42.

# Chapter 15

# Gradualism *versus* the Fossil Record

The term fossil record refers to the remains that we find today, including bones, shells, and footprints, of ancient creatures that long ago inhabited the Earth and have since become extinct. The question before us relates to the *rate* at which these ancient creatures underwent evolutionary changes to become the animals of today.

Did evolutionary changes occur *slowly and gradually*, over many thousands of years, or did evolutionary changes occur *relatively rapidly*? We shall here discuss which of these two possible rates of evolutionary change is consistent with the fossil record.

## Richard Dawkins

Evolutionary biologist Richard Dawkins, of the University of Oxford, made the following statement about the role of gradualism in evolution (*emphasis added*)[1]:

> "Cumulative selection, *by slow and gradual degrees,* is the explanation for life's complex design ... *slow, gradual, cumulative natural selection* is the ultimate explanation of our existence ... *Gradualism is the very heart* of evolution."

We see the uncompromising view of Dawkins that the fossil record supports gradualism.

# Stephen Jay Gould

Evolutionary biologist Stephen Jay Gould, of Harvard University, made a very different statement regarding gradualism in the fossil record (*emphasis added*)[2]:

> "The history of most fossil species includes features *particularly inconsistent with gradualism:* (i) Statis: Most species exhibit no directional change during their tenure on Earth. They appear in the fossil record *looking much the same* as when they disappear. Morphological change is usually *limited and directionless.*
>
> (ii) Sudden appearance: *A species does not arise by the gradual transformation of its ancestors.* It appears in the fossil record *all at once and fully formed.*"

# An Astonishing Situation

The two above-mentioned respected evolutionary biologists looked at the same fossil data and came to completely opposite conclusions! It is clear from these two contradictory assertions that the role of gradualism in evolution is far from simple.

# Punctuated Equilibrium

In 1972, on the basis of a detailed study of the fossil record, Gould and his colleague, evolutionary biologist Niles Eldredge, introduced the concept of punctuated equilibrium.[3] They reported that their study of fossils indicated that the *gradual evolution of species does not characterize the fossil record.* Instead, they found that a species typically remains almost unchanged, in a state of *"equilibrium,"* for long periods of time, and then, this equilibrium is suddenly *"punctuated"* by rapid evolutionary change.

This absence of gradualism refers to the *major taxonomic groups*, at the level of the phylum, class, and order. At the *much lower taxonomic level* of the genus and species, there are many examples of gradualism. For example, wolves evolved into dogs. However, the main task of a theory of evolution is to explain the *large changes* in the animal kingdom, the appearance of new orders, new classes, and new phyla.

The concept of punctuated equilibrium is an approach to evolution that is opposed to the gradualism that is so strongly emphasized by Richard Dawkins. It is therefore not surprising that Dawkins devoted an entire chapter of his book, *The Blind Watchmaker*, to attacking that concept.[4] Dawkins tries to trivialize the idea, asserting that the concept of punctuated equilibrium is utterly devoid of any significance. He writes time after time that "punctuated equilibrium is but a minor wrinkle on the surface of neo-Darwinian theory" and a "minor gloss on Darwinism" and that "it is a minor variety of Darwinism."

In view of these two conflicting views, it is instructive to read the opinions of other authorities regarding the concept of punctuated equilibrium.

*The Cambridge Encyclopedia of Earth Sciences* considers punctuated equilibrium sufficiently important to warrant a detailed discussion, in which the following is asserted[5]:

> *"The model of punctuated equilibrium has recently become widely accepted."*

The journal *Europhysics News* reports the following[6]:

> *"One characteristic of the fossil record is the presence of 'punctuations,' where new species appear on a relatively short geological time scale. Evolution is not a slow process, whereby one species gradually changes into another."*

In fact, every modern textbook on biology discusses punctuated equilibrium.[7]

Steven Stanley, of Johns Hopkins University, a leading authority on evolution, considered the concept of punctuated equilibrium so significant that he wrote a book to explain to the layman this new approach to evolutionary biology in which he stated the following[8]:

> *"What I shall describe in this book is evidence that evolution is not quite what nearly all of us thought it to be a decade ago. The evidence comes from the record of fossils — a record which now reveals that most evolution takes place rapidly.*

> *While this "punctuational" view has displaced the traditional "gradualistic" view in the minds of many evolutionists, there remain dissenters. In this book, I offer opposition to the traditional portrayal. I give the interested non-specialist access to the punctuational view and its implications ... The emergence of the punctuational model of evolution is an exciting time in evolutionary science."*

Note the striking differences between these two views. Whereas Stanley and other quoted authorities stress the *importance* of the concept of punctuated equilibrium (*"an exciting time in evolutionary science"*), Dawkins simply *dismisses* the concept (*"minor wrinkle"*).

## Ancient Fossils

It is interesting to examine the evolution of the earliest living creatures in order to learn if they show signs of gradualism. The Earth was formed about 4.5 billion years ago. The surface of the Earth was initially too hot for solid rocks to form. The earliest solid rocks date from about four billion years ago. Even in these most ancient rocks, one finds signs of life in the form of *bacteria*, the simplest living creatures. (We shall not discuss whether viruses are living creatures.)

The next step in animal diversity was the evolution of *one-celled protozoa*, such as the amoebae. The amoebae, though only a fraction of a millimeter in size, are far larger and vastly more complicated than any bacterium. It required *nearly three billion years* (!) until some bacteria evolved into one-celled protozoa[9]:

> *"Nearly two thirds of life's history belongs exclusively to bacteria and blue-green algae that look pretty much the same at the end of these three billion years as at the beginning."*

The next step in animal diversity was the evolution of *simple multicellular creatures*. After *nearly a billion years* (!), simple multicellular creatures appeared.

We are now about 600 million years in the past, at the beginning of the Cambrian Period. This is the earliest period in which multicellular animals appeared. The next step in animal diversity was the evolution of the antecedents of the animals that we observe today. This step occurred so rapidly that evolutionary biologists speak of an *"explosion of life-forms."* The following quotes will demonstrate the rapidity of this step:

> *"Perhaps the most astonishing aspect of the Cambrian fauna is that so many radically different types of animals appeared in such a short interval ... The Cambrian Period saw the appearance of new phyla and classes of animals at a rate that has not been matched since ... A tremendous explosion of life took place. The Cambrian marks the first appearance of many major groups of animals."*[10]
>
> *"The Cambrian period ushered in an explosion of multicellular lifeforms ... the explosion of lifeforms which inaugurated the Cambrian period."*[11]
>
> *"A hypothetical observer in the late Precambrian Period would have had few grounds for optimism about the future of life. Yet, over a comparatively short period ... the initial diversification of metazoans [animals] gave way to a series of dazzling radiations during the Cambrian period ... Paleontologists are impressed by the rapidity of the development of such a diverse range of organisms, including ten or more invertebrate phyla."*[12]

Scientists have long sought — without much success — to understand this *"explosion of lifeforms,"* characterized by the *"sudden appearance"* and *"dazzling radiation"* of so many different types of animals, all dating from about the same time.

## The Plant Kingdom

We have so far discussed the *"evolutionary explosion"* of animals that occurred about 600 million years ago, in the Cambrian Period. What about the plant kingdom? Does the fossil evidence of ancient plants support gradualism?

Most evolutionary biologists study the animal kingdom. However, the evolution of the plant kingdom has also been investigated. The fossil evidence shows that the plant kingdom also experienced an *"evolutionary explosion*" quite similar to that experienced by the animal kingdom. The "*evolutionary explosion*" of plants occurred in the Devonian Period, about 390 million years ago, and is known as the *Devonian explosion of plant life*.

## Conclusion

The fossil evidence indicates that neither the animal kingdom nor the plant kingdom evolved gradually. Punctuated equilibrium is a much more accurate description of the fossil record.

## References

1. R. Dawkins, 1986, *The Blind Watchmaker* (W. W. Norton: London and New York), pp. 317–318.
2. S. J. Gould, 1980, *The Panda's Thumb* (W. W. Norton: London and New York), p. 151.
3. S. J. Gould and N. Eldredge, 1972, in *Models in Paleobiology*, editor T. J. M. Schopf (Freeman Cooper: San Francisco), pp. 82–115.
4. R. Dawkins, 1986, *The Blind Watchmaker* (W. W. Norton: London and New York), pp. 250, 251, 287.
5. D. J. Smith, editor, 1981, *The Cambridge Encyclopedia of Earth Sciences* (Cambridge University Press: Cambridge), p. 381; see also *new edition*, 2002, (Cambridge University Press: Cambridge), p. 538.
6. P. Alstrom, 1999, *Europhysics News*, vol. 30, p. 22.
7. See, for example, the widely used university textbook, E. P. Solomon *et al.*, 1996, *Biology*, fourth edition (Hartcourt Brace College Publishers: New York), p. 442.
8. S. M. Stanley, 1981, *The New Evolutionary Timetable* (Basic Books: New York), pp. xv–xvi.
9. S. J. Gould, 1987, *An Urchin in the Storm* (Penguin Books: London), p. 212.
10. M. A. S. McMenamin, April 1987, *Scientific American*, pp. 84, 90.
11. F. H. Shu, 1982, *The Physical Universe* (University Science Books: Mill Valley, California), pp. 503, 545.
12. D. J. Smith, editor, 1981, *The Cambridge Encyclopedia of Earth Sciences* (Cambridge University Press: Cambridge), pp. 370–371.

# Chapter 16

# Mass Extinctions: *"Bad Genes or Bad Luck?"*

Our knowledge of the ancient animal kingdom comes primarily from fossil evidence, the remains of ancient creatures. Although there is also another method for determining evolutionary history, based on DNA studies, fossil evidence retains its importance. The most astonishing phenomena regarding the fossil evidence are *mass extinctions*. A mass extinction is a catastrophic event that leads to *the extinction of the majority of the world's species*. The destroyed species never reappear in the fossil record. The geological time scale is traditionally separated into Eras, with each Era ending with the occurrence of a mass extinction.

The largest mass extinction, in which over 90% of all species were destroyed, marks the boundary between the Paleozoic Era and the Mesozoic Era. The most famous mass extinction relates to the sudden disappearance of all the world's dinosaurs. This mass extinction marks the boundary between the Mesozoic Era and the Cenozoic Era. Here, we will concentrate on the sudden extinction of all the world's dinosaurs in the mass extinction that ended the Mesozoic Era.

## Dinosaurs

The dinosaurs were one of the most successful groups of animals that ever lived. Dinosaurs and their close relatives lived on every continent, in the air (flying dinosaurs), and in all the oceans (marine dinosaurs). For nearly

150 million years, dinosaurs flourished. Other animals lived in constant danger of being eaten or destroyed by these terrible monsters. (The word "dinosaur" means "terrible lizard.") Because of the dominance of the dinosaurs, this era is commonly called the Age of Reptiles.

About 65 million years ago, all dinosaurs disappeared from the fossil record. Together with the dinosaurs, over 70% of all the world's species suddenly became extinct. What catastrophic event led to the sudden destruction of the highly successful dinosaurs that had flourished for such a long time? Many proposals were put forward to explain this mass extinction, but none were convincing.

## The Alvarez Proposal

In 1980, Nobel laureate Luis Alvarez and his son Walter presented a radically different proposal. They suggested that the mass extinction was caused by a large meteor about the size of Singapore that collided with the Earth. The collision pulverized the meteor, and the dust and debris that were thrown up blocked the sunlight worldwide, destroying most plant life. The dust also poisoned the atmosphere, with strong winds spreading the poisonous fumes worldwide. In addition, the rapid passage of the meteor through the atmosphere caused enormous fires. The combination of all these catastrophes caused the mass extinction.

The Alvarez idea met with widespread skepticism. It didn't help that Luis Alvarez wasn't even a biologist. He had received his Nobel Prize for his physics research. However, the Alvarez team patiently gathered the evidence. It was found that the impact of the meteor had occurred in the Yucatan Peninsula in Mexico. In the July 1987 issue of *Physics Today*, Luis Alvarez published an article in which he presented *fifteen different pieces of scientific evidence* to support his proposal.

The most compelling evidence was the thin layer of the metal iridium that was found worldwide just below the surface of the Earth. Ordinarily, the metal iridium is found *far below* the surface of the Earth, with only very insignificant amounts found *near* the surface. The concentration of iridium found near the surface was *hundreds of times* greater than expected. Where did this unexpectedly large amount of iridium come from? Since a meteor is rich in iridium, the pulverized meteor explains the extremely large concentration of the metal iridium found near the surface.

Although a few dissenting voices can still be heard, the Alvarez proposal has become widely accepted as the correct explanation for this mass extinction.

## Relevance of Mass Extinctions to Evolution

What is important to our discussion is the relationship between mass extinctions and Darwin's theory of evolution. According to Darwin's principle of "survival of the fittest," the individuals that manage to survive the "struggle for existence" possess favorable genetic traits that make them more fit. These traits include being stronger, faster, and less susceptible to disease than their competitors. In informal language, the survivors have "good genes." The individuals that do not survive are those that lack such favorable traits. In other words, the losers in the "struggle for existence" had "bad genes."

When the meteor fell from the sky and destroyed all the dinosaurs, the dinosaurs that died were not lacking favorable genetic traits. They did not *suddenly* become weaker or slower or more susceptible to disease. Their extinction was due to extraterrestrial circumstances — a meteor from outer space. The dinosaurs simply had the *bad luck* of "being in the wrong place at the wrong time." The surviving species did not suddenly develop more favorable genetic traits. The survivors simply had *better luck* than the dinosaurs in escaping this worldwide catastrophe.

The idea of bad luck causing this mass extinction was the subject of an article, later expanded into a book, by David Raup, former president of the American Paleontological Union, entitled *Extinctions: Bad Genes or Bad Luck?* In Raup's title, *"Bad Genes"* relates to "Darwin's theory," whereas *"Bad Luck"* relates to "the Alvarez theory." Raup points out that a mass extinction is not explained by Darwin's theory[1]:

> *"If the extinction of a given species is more bad luck than bad genes, then the conventional Darwinian model is not relevant ... Pure chance*

*would favor some biologic groups over others — all in the absence of Darwinian natural selection."*

The important role played by luck in mass extinctions has been emphasized by many evolutionary biologists, including Stephen Jay Gould, of Harvard University[2]:

*"Many people feel that conventional Darwinian reasons must rule the ebb and flow of major groups ... But recent data on the extent of a mass extinction calls this comforting explanation into question. The survivors may simply be among the lucky ones ... evolutionary theory is stirring from the strict Darwinism that previously prevailed ... This is an exciting development in evolutionary theory."*

George Yule, of the University of Oxford, expressed this idea in the following way[3]:

*"The species that were exterminated were not killed because of inherent defects, but simply because they had the ill-luck to stand in the way of the cataclysm."*

David Jablonski, of the University of Chicago, an authority on the subject of mass extinctions, stated the following[4]:

*"When a mass extinction strikes, it is not the "most fit" species that survive; it is the most fortunate. Species that had been barely hanging on, suddenly inherit the Earth."*

These leading paleontologists all emphasize that if a large meteor suddenly falls from the sky, wiping out some species while permitting other species to survive and ultimately to flourish, then the Darwinian principle of survival of the fittest is irrelevant. The species that survived were simply blessed with *good luck* — survival upon the occurrence of an extremely improbable and totally unexpected event.

## Dinosaurs and Man

There is an important relationship between dinosaurs and human beings. As long as dinosaurs dominated the Earth, there was no possibility for large mammals to evolve (human beings are large mammals). Only after the dinosaurs were wiped out could mammals flourish and grow in size.

The important connection between human beings and the dinosaurs was pointed out by Alvarez, who ends his article regarding the meteoric impact that destroyed all the dinosaurs with the following words[5]:

> *"From our human point of view, that impact was one of the most important events in the history of our planet. Had it not taken place, the largest mammals alive today might still resemble the small rat-like creatures that were then scurrying around trying to avoid being destroyed by the dinosaurs."*

## References

1. D. M. Raup, 1981, *Acta Geologica Hispanica*, vol. 15, pp. 25–33.
2. S. J. Gould, 1983, *Hen's Teeth and Horse's Toes* (W. W. Norton: London and New York), pp. 320–324.
3. G. Yule, 1979, *Philosophical Transaction of the Royal Society*, vol. 213, p. 24.
4. D. Jablonski, June 1989, quoted in *National Geographic*, p. 673.
5. L. W. Alvarez, July 1987, *Physics Today*, pp. 24–33.

# Chapter 17

# Darwin's Theory of Evolution: Pros and Cons

Darwin's theory of evolution states the following. Different individuals of a given species have different physical characteristics and, as a result, some individuals are better able to survive ("*more fit*") than others. In each generation, many more offspring are produced than can possibly survive to the next generation. The food supply is limited. Moreover, many individual animals are destroyed by predators or die from disease. This is called the "*struggle for existence*." The individuals that survive long enough to reproduce the next generation are those having characteristics that enhance their survival. These include being stronger or faster or less susceptible to disease than other individuals of the same species. The process by which nature weeds out the "*less fit*" is called "*natural selection*." In the course of very many generations, the *more fit* animals become dominant. The characteristics that favor survival become incorporated into the animal's gene pool. Eventually, there are so many changes in the gene pool that a new species is formed. In this way, new species gradually evolve. This, in a nutshell, is Darwin's theory of evolution.

## Pro: Richard Dawkins

One of Darwin's greatest champions is Richard Dawkins of the University of Oxford. In his book, *The Blind Watchmaker*, Dawkins describes Darwin's theory in the following manner[1]: "*the elegant and beautiful solution to this deepest of problems [evolution]."* Since Darwin's theory is *"a remarkably simple theory,"* Dawkins asks, *"How could such a simple idea go so long undiscovered by great thinkers?"* But we shall see that there are many outstanding evolutionary biologists who question Darwin's theory.

What did Dawkins write about the important topic of mass extinctions? When a mass extinction occurs, a majority of the world's species are suddenly destroyed within *a relatively short time*. Thus, mass extinctions would seem to pose a significant challenge to the gradual evolution proposed by Darwin and championed by Dawkins. How did Dawkins handle this challenge?

*Dawkins handled this challenge by simply ignoring mass extinctions. They are never mentioned in his book*!

The importance of mass extinctions in the role of evolution has been emphasized by leading scientists.

Nobel laureate Luis Alverez writes the following[2]:

*"That impact [of the meteor that caused the mass extinction of the dinosaurs and 70% of all other species] was one of the most important events in the history of our planet.*

David Raup, former president of the American Paleontological Union, writes the following[3]:

*"If the extinction of a given species is more bad luck than bad genes, then the conventional Darwinian model is not relevant."*

## Pro: Daniel Dennett

Another of Darwin's champions is Daniel Dennett, Distinguished Professor of Arts and Sciences at Tufts University in Medford, Massachusetts. Dennett published a book in praise of Darwin's theory, writing the following[4]:

*"If I were to give an award for the single best idea that anyone ever had, I'd give it to Darwin, ahead of Newton and Einstein and everyone else."*

Rare praise indeed, since Newton and Einstein are universally considered to be the two greatest physicists of all time. Dennett explains the reason for his unbounded admiration for "Darwin's magnificent idea":

*"Darwin's idea had been born to answer questions in biology, but it also offers answers to questions in cosmology and psychology ... And if mindless evolution could account for the breathtakingly clever artifacts of the biosphere, how could the products of our minds be exempt?"*

Dennett thus writes that Darwin's idea is not limited to explaining evolution, but it also answers fundamental questions in subjects as far afield as cosmology and psychology, including the very workings of the human mind.

In view of Dennett's far-reaching praise, it is surprising to learn that there are respected evolutionary biologists (to be listed in this chapter) who seriously doubt whether Darwin's theory explains even the evolution of the animal kingdom.

What did Dennett's colleagues think of his book? Evolutionary biologist H. Allen Orr, of the University of Rochester, published a devastating review of Dennett's book in a major journal in evolutionary biology in which he stated the following[5]:

*"Dennett's book suffers from a number of problems ... marred by factual errors ... undermining any hope of a balanced presentation ... fails to appreciate concerns ... fundamentally misunderstands ... is obsessed with defending adaptive story-telling within biology ... never confronts legitimate worries ... easier for him to ridicule ... betrays fundamental errors ... confusing about how far Darwinism extends ... chief claim is unconvincing ... evidence for each claim is non-existent."*

Stephen Jay Gould, of Harvard University, characterized Dennett's book as follows[6]:

*"fallacy ... discussions resting on ridicule and error ... false attribution ... misreadings ... false charges ... gratuitous speculation of motives ... high density of errors ... simplistic distortion of Darwin's theory."*

# Con: Steven Stanley

Steven Stanley, of Johns Hopkins University, another leading authority, published a book to explain the recent non-Darwinian approach to evolution, writing the following[7]:

> *"Darwin and the many architects of neo-Darwinism would have been confounded by the fossil evidence ... they would have been shocked ... It was a gradualistic view of evolution that led Darwin and the others on their fruitless search ... Believing only in slow, persistent, gradual evolution, they postulated an undocumented, incorrect history of early animal life ... The fossil record has now answered with solid evidence, confronting gradualism with an insoluble problem ... Darwin reached in desperation for an ancient mythical kingdom ... Other workers sustained this fanciful history into the 1960s to permit gradual evolution to account for the complexity of flowering plants ... Such assignments have now all been deemed erroneous."*

# Con: Kenneth Hsu

Kenneth Hsu, of the Swiss Institute of Technology (ETH), points out that important evidence that many lifeforms *suddenly* became extinct comes from stratigraphic studies of magnetic sediments on the ocean floor. Hsu writes as follows[8]:

> *"Magnetostratigraphic investigations during the last few years have certainly dealt the final blow to Darwin's postulate."*

# Non-Darwinian Theories of Evolution

Several non-Darwinian theories have been proposed to account for the vast panorama of animal and plant life that we observe today. Two of these non-Darwinian theories will now be discussed.

These theories of evolution are non-Darwinian in the sense that the explanations they propose for biological evolution are *unrelated* to the basic Darwinian ideas of survival of the fittest, adaptation, struggle for existence, and natural selection.

# Complexity Theory

One of the most exciting discoveries in recent years is known as "complexity theory" (also called "self-organized criticality"). Complexity theory scientists have found that it is characteristic of certain complex systems to remain static for a long period of time and then *suddenly* undergo a rapid fundamental change, *spontaneously, without any driving force.*[9] The implications of complexity theory are widespread, including the physical sciences, the life sciences, and even the social sciences. For example, it is now widely thought that the underlying cause of the stock market crash of 1987 was not connected to economics. Rather, the market crashed *spontaneously*, because of the complexity of the system of stock trading.

The new findings of complexity theory were deemed so important that in 1984, George Cowan and three Nobel laureates (Philip Anderson, Murray Gell-Mann, and Kenneth Arrow) established the Santa Fe Institute in New Mexico as an interdisciplinary research institution devoted to the study of the many implications of complexity theory.

(An explanation of the remarkable behavior of certain complex systems is far beyond the scope of this book. Even simplified "popular" versions of complexity theory present a daunting experience for the layman. But for those who like a challenge, introductions to complexity theory are given in *How Nature Works* by Per Bak[10] and in *At Home in the Universe: Laws of Self-Organization and Complexity* by Stuart Kauffman.[11] Both authors are among the founders of this new science.)

# Complexity Theory and Evolution

Scientists have recognized that living systems — animals and plants — satisfy the criteria of complexity theory. Moreover, the fundamental changes that have occurred over time in living creatures — evolution — are very reminiscent of the *spontaneous* changes that characterize certain complex systems. Therefore, it has been suggested that biological evolution *may not be due* to the Darwinian mechanisms of survival of the fittest and adaptation. Rather, the behavior of complex systems may be the correct explanation for the evolutionary changes observed in the animal kingdom.

The most striking similarity between biological evolution and complexity theory lies in the phenomenon known as punctuated equilibrium. This concept expresses the fact that the fossil record indicates that most evolutionary changes have occurred *suddenly*, exactly as predicted by complexity theory, rather than gradually, as predicted by Darwin's theory.

It is instructive to quote leading scientists on the relationship between Darwin's theory and complexity theory.

According to Stuart Kauffman, of the Santa Fe Institute,[12]

> *"We live in a world of stunning biological complexity. Where did this grand architecture come from? Darwin taught us that the order of the biological world arose as natural selection sifted among random mutations for the rare, useful forms. Recent research has shown that this dominant view of biology is not complete. Self-organization is the root source of order."*

Henrik Jensen, of Imperial College, London, shares this view. He emphasizes that many features of the evolutionary development of the animal kingdom are exhibited by complex systems[13]:

> *"Since the 1970s, it has been known that evolution takes place through bursts of activity, separated by calm periods ... Species survive for long periods of time, and then disappear within a relatively short span of years ... When evolution and extinction are portrayed in this fashion, we immediately spot the similarities with the dynamics of complex systems."*

## The Neutral Theory of Molecular Evolution

Another important non-Darwinian theory of evolution is the neutral theory of molecular evolution. This theory was proposed in 1968 by leading geneticist Motoo Kimura of the National Institute of Genetics in Japan.[14] Kimura asserts that the genetic mutations that become part of the gene pool are "selectively neutral," meaning that they are *neither more nor less advantageous* than the genes they replace[15]:

> *"For more than a decade, I have championed a view different from Darwin ... My theory holds that at the molecular level, most evolutionary change and most of the variability within a species are not caused by Darwinian selection, but by the 'random drift' of mutant genes that are selectively neutral ... The picture of evolutionary change that emerged from molecular studies seems to be quite incompatible with the expectations of neo-Darwinism."*

## Scientific Revolutions

The reluctance of some biologists to admit that a new paradigm has occurred in evolutionary biology is an example of the phenomenon described by the philosopher of science Thomas Kuhn in his important book, *The Structure of Scientific Revolutions.*[16] Kuhn points out that when a scientific revolution takes place, leading to a new paradigm, many scientists find themselves unable to accept the new results. They continue to insist that the new revolutionary ideas "fit under the rubric" of the older ideas, and that no change in thinking is required.

## References

1. R. Dawkins, 1986, *The Blind Watchmaker* (W. W. Norton: London and New York), pp. ix–xi.
2. L. W. Alvarez, July 1987, *Physics Today*, p. 33.
3. D. M. Raup, 1981, *Acta Geologica Hispanica*, vol. 15, p. 30.
4. D. C. Dennett, 1995, *Darwin's Dangerous Idea* (Penguin Books: London and New York), pp. 21, 63.
5. H. A. Orr, 1996, *Evolution*, vol. 50, pp. 467–472.
6. S. J. Gould, 12 June 1997, *The New York Review*, p. 36.
7. S. M. Stanley, *The New Evolutionary Timetable*, 1981 (Basic Books: New York), p. xv.

*The Wonders of Science*

8. K. J. Hsu, 1986, *The Great Dying* (Harcourt Brace: New York), p. 88.
9. P. Bak *et al.*, 1987, *Physical Review Letters*, vol. 59, pp. 381–384.
10. P. Bak, 1996, *How Nature Works* (Springer-Verlag: New York).
11. S. A. Kauffman, 1995, *At Home in the Universe: the Laws of Self-Organization and Complexity* (Oxford University Press: Oxford).
12. S. A. Kauffman, 1995, *At Home in the Universe: the Laws of Self-Organization* and Complexity (Oxford University Press), pp. vii–viii, 8.
13. H. J. Jensen, 1998, *Self-Organized Criticality: Emergent Complex Behavior in Biological Systems* (Cambridge University Press, Cambridge), p. 27.
14. M. Kimura, February 1968, *Nature*, pp. 624–626.
15. M. Kimura, November 1979, *Scientific American*, pp. 94, 98, 101.
16. T. S. Kuhn, 1962, *The Structure of Scientific Revolutions* (University of Chicago Press: Chicago).

# Chapter 18

# Darwin's Theory of Evolution: Confusion

Nearly two centuries have passed since Charles Darwin published his book, *The Origin of Species*, in 1859. Almost a century has passed since the formulation of neo-Darwinism, the modern version of Darwin's theory of evolution. The modern theory is also known as the "synthetic theory of evolution," because it is a *synthesis* of Darwin's original ideas (natural selection, survival of the fittest, struggle for existence, and adaptation) with genetics. During this period, Darwin's theory of evolution became one of the best-known theories in the realm of science. This theory is taught in every university; it is discussed by every educated person; it has been the subject of innumerable articles. It is difficult to believe that there is anyone who has not heard of Darwin's theory.

In view of this long history, one would expect to find overall agreement among evolutionary biologists regarding the content of Darwin's theory and the basic facts about evolution. In science, unlike other disciplines, there are objective methods of examining data and arriving at conclusions that are accepted by the entire scientific community. Therefore, one would have thought that scientists have long since sorted out the basic principles of evolution and Darwin's theory. But this is by no means the case. Strident arguments persist to this day among leading authorities concerning even the most fundamental principles of evolution. The controversies deal with fundamental issues that have been debated by evolutionary biologists for decades without having achieved a scientific consensus.

## "The Species Problem"

Biologists are still debating about what constitutes a species or even whether the concept of "species" has any meaning. The controversy regarding this question has persisted for decades and has been given a formal name: *the species problem.*" The first two papers in the *Proceedings* of the First International Symposium on Biology presented the opposing views:

According to Peter Raven, Director of the Missouri Botanical Gardens,[1]

> *"One should turn away from the biological species as a unit of funda-mental evolutionary significance ... species do not have an objective reality in nature ... It is the population, not the species, which should be the focus of evolutionary studies ... We have embarrassingly little evi-dence for some widely-held notions in systematics and evolutionary biology."*

In complete contrast to this view, Walter Bock, of Columbia University, writes,[2]

> *"Species have an objective reality ... Species and speciation play a very important role in macro-evolutionary change ... Interactions between species provide the major source of directional selective pressures that drive evolutionary changes."*

## What is a Species?

The standard definition of a species is a group of animals that form an interbreeding population. In other words, the standard criterion for deter-mining whether or not two animals belong to the same species is whether or not they successfully interbreed in the wild to produce fertile offspring.

(In a zoo, one can artificially cause interbreeding among different species — lions and tigers, for example.)

Unfortunately, this standard definition of a species has been found to be incorrect. It is now recognized that there are many examples of animals that clearly belong to different species, but nevertheless interbreed quite freely in the wild to produce fertile hybrids.

James Mallet, of University College, London, writes,[3]

> "About a quarter of all species of ducks, game birds, and pheasants hybridize [interspecies breeding] in the wild ... For coral reef fish, the figure is 20% ... Blue whales hybridize with fin whales ... A quarter of all Heliconius species of butterflies hybridize in the wild."

## Are Mutations Random?

A creature is determined by its array of genes, which are segments of the long thread-like molecules of DNA found in every cell of every living creature. Elephants have elephant genes; cockroaches have cockroach genes; wheat has wheat genes. During sexual reproduction, the genes of the parents are copied to form the genes that are transmitted to their offspring. In the process of copying the genes, an error — *mutation* — may occasionally occur. The resulting offspring is said to have undergone a genetic mutation.

One of the cornerstones of evolutionary biology is that all mutations are *random*. Mutations are not *directed* to solve some problem faced by the organism. There is no connection whatsoever between the occurrence of a mutation and its being beneficial or harmful to the organism. Mutations are always the result of *pure chance*.

Random mutations constitute the raw material on which natural selection works. Darwin's theory of evolution is based on the twin pillars of *random* mutations, which form genetic variety among a population of animals, and *non-random* natural selection, which "selects" the individuals that are more fit to survive. Thus, evolution is driven by the *interplay* between random processes (mutations) and non-random processes (natural selection). This is an essential feature of Darwin's theory of evolution.

In view of this fundamental understanding of Darwin's theory, it came as quite a surprise when some experiments suggested that mutations may not be random after all. John Cairns and his colleagues, of the Harvard University School of Public Health, reported in the influential journal *Nature* "*that cells may possess mechanisms for choosing which mutations will occur.*"

In their experiment, Cairns and co-workers developed a strain of bacteria that could *not* digest lactose (milk sugar) and then fed them *only* lactose. In the absence of food that they could digest, these bacteria were not expected to multiply. But instead of remaining inert, some of these bacteria *mutated* one of their genes to make the enzyme that *permitted* the digestion of lactose. These mutated bacteria then flourished.

The researchers were astonished by what they had discovered ("*That such events ever occur seems almost unbelievable*"). They emphasized the far-reaching implications[4]:

> "*The main purpose of this paper is to show how insecure is our belief in the randomness of mutations. This idea [mutations are random] seems to be a doctrine that has never been properly put to the test. We describe experiments and evidence suggesting that bacteria can choose the mutations they should produce.*"

The revolutionary nature of these findings, termed *directed mutations*, was emphasized by *Scientific American*. This journal called these results "*sensational experiments*" leading to an "*incendiary idea.*" *Scientific American* entitled their article "Evolution Evolving," with the subtitle "New findings suggest that mutation is more complicated than anyone thought." The article proclaimed that the experimental results of the Harvard University team contradicted fundamental principles of evolutionary biology[5]:

> "*This radical proposal [directed mutations] collided head-on with the sacrosanct principle that mutations occur at a rate that is completely unrelated to whatever consequences they might have for the organism ... This incendiary idea ignited a firestorm of debate.*"

The idea of directed mutation is so unthinkable that in the years following the Cairns experiments, various explanations for these results were proposed. Barry Hall, of Rochester University, writes that whatever explanation finally does solve this puzzle, *"the mechanisms that produce the ghost of directed mutation could shake up biology."*[6]

## What Factors Control Evolution?

Biologists are still arguing about what factors control evolution. For example, the editors of *Causes of Evolution* explain the purpose of their volume as follows[7]:

*"We wanted to see what would happen if paleobiologists were asked to identify what they believed were the causal factors controlling evolution ... Identifying the causal factors proposed by evolutionary biologists could better contribute to understanding of the role that different factors play in evolutionary processes."*

The centrality of this question to evolutionary biology was noted by Stephen Jay Gould of Harvard University[8]:

*"The two contradictory dichotomies selected by the editors have been central to evolutionary biology ever since Darwin — and have never resolved."*

Gould's assessment was shared by Norman Gilinsky of Virginia State University[9]:

*"When discussion and debate over an issue continue for over a century with little prospect for resolution, it's time to re-evaluate the question ... This emphasizes the need to rethink our entire conceptualization of organism and environment and, thereby, the entire question of biotic versus abiotic causes."*

It is surprising to read in the twenty-first century of *"the need to rethink our entire conceptualization"* of a problem that *"has been central to evolutionary biology ever since Darwin."*

## Has There Been Any Evolutionary Progress in the Animal Kingdom?

Darwin's motivation for proposing his theory was to suggest a *mechanism* that would explain *how* progressive biological change came about. Therefore, it is surprising to read that there is a dispute regarding whether evolutionary progress has occurred. But dispute there is!

Richard Dawkins and Stephen Jay Gould are respected evolutionary biologists. Both are professors at important universities (Dawkins at Oxford; Gould at Harvard). Both have introduced major concepts into the field of evolutionary biology (Dawkins: the selfish gene; Gould: punctuated equilibrium). Both are eloquent and prolific writers who have authored best-sellers on evolutionary biology. But here the similarity ends. These two biologists *completely disagree* on whether evolutionary progress has occurred in the animal kingdom. Dawkins answers with an unequivocal "Yes!" and Gould answers with an equally unequivocal "No!"

To stimulate debate, the editors of *Evolution*, the premier scientific journal in the field, decided to pit these two evolutionary biologists against each other, by having each man review the other's recent book (*Climbing Mount Improbable* by Dawkins; *Life's Grandeur* by Gould). The editors noted with satisfaction that "*we expected — and received — two frank and provocative articles about important issues.*"

The most important issue was whether or not there has been evolutionary progress over the history of the animal kingdom. Gould maintains that "*the notion of 'progress' in evolution is an illusion.*" However, Dawkins maintains that "*adaptive evolution is deeply and indispensably progressive.*"

This controversy shows how two respected scientists (Dawkins and Gould) can look at the same data and, nevertheless, come to opposite conclusions! The editors of the important journal *Evolution* write the following[10]:

> "*Throughout the past half century, our journal has witnessed many controversies, but none more persistent than the debate about the roles of*

*random versus deterministic forces in evolution. Stephen Jay Gould has constantly emphasized the importance of chance and contingency, whereas Richard Dawkins embraces a deterministic world view, emphasizing the power of natural selection."*

Furthermore, each of these two famous evolutionary biologists claims that his equally famous colleague does not understand the most basic principles of evolutionary biology!

Dawkins writes of Gould, *"His argument is flawed ... Gould is wrong ... Gould's attempt to reduce progress to a trivial artifact constitutes a surprising impoverishment, and an unwonted demeaning of the richness of evolutionary processes."*

Gould writes of Dawkins, *"His logic is flawed by an ill-chosen metaphorical apparatus ... such imagery is especially misleading ... the conceptual foundation of Dawkins's book is deeply invalid at its core."*

It should be emphasized that Gould and Dawkins are arguing about the very essence of their discipline — matters that have been discussed in detail for many decades. In evolutionary biology, one finds that even the fundamental principles are still the subject of sharp dispute.

## Natural Selection

John Endler, of the University of California, stated the following[11]:

> *"Those who think that natural selection is the most important factor in evolution, work primarily with morphological traits in natural populations, whereas those who consider natural selection to be unimportant tend to work with molecular or biochemical traits in laboratory populations."*

Jeffry Mitton, of the University of Colorado, said the following[12]:

> *"We find that natural selection means different things to different people ... Differences of opinion concerning the nature and importance of natural selection contribute to several unresolved issues."*

# Speciation

Bradley Shaffer, of the University of California, laments the scientific controversies associated with speciation, which is one of the central issues of evolutionary biology[13]:

> *"Speciation. Modes of speciation. Mechanisms, hybrid zones, isolating mechanisms, reinforcement. These words, and many others associated with speciation research, often bring fear to practicing evolutionary biologists, and certainly to us who try to teach evolution. No one seems to be able to agree even on the terminology associated with speciation, let alone the proper way to study the process ... What an unfortunate way to deal with a topic that has been a central focus of evolutionary biology ever since Darwin."*

# Conclusion

One notes the confusion that abounds in evolutionary biology. Although this field has been the subject of intensive research for many decades, scientific consensus regarding the basic principles has not yet materialized.

# References

1. P. H. Raven, 1986, in *Modern Aspects of Species*, editor K. Iwatsuki (University of Tokyo Press: Tokyo), pp. 11–29.
2. W. J. Bock, 1986, in *Modern Aspects of Species*, editor K. Iwatsuki (University of Tokyo Press: Tokyo), pp. 31–57.
3. J. Mallet, 3 July 1999, *New Scientist*, pp. 32–36.
4. J. Cairns *et al.*, 1988, *Nature*, vol. 335, pp. 142–145.
5. "In Focus", September 1997, *Scientific American*, p. 9.
6. B. G. Hall, September 1997, *Scientific American*, p. 12.
7. R. Ross and W. Allmon, editors, 1990, *Causes of Evolution* (University of Chicago Press: Chicago), p. 1.
8. S. J. Gould, 1990, in *Causes of Evolution*, editors R. Ross and W. Allmon (University of Chicago Press: Chicago), p. vii.
9. N. L. Gilinsky, 1992, *Evolution*, vol. 46, pp. 578–579.

10. J. Coyne, 1997, *Evolution*, vol. 51, p. 1015.
11. J. A. Endler, 1986, *Natural Selection in the Wild* (Princeton University Press: Princeton), p. 9.
12. J. B. Mitton, 1989, *Evolution*, vol. 43, p. 1339.
13. H. B. Shaffer, 1990, *Evolution*, vol. 44, p. 1711.

# Chapter 19

# Evolution of Human Beings

## Paleoanthropology

The most fascinating topic in fossil research is the study of the origins of *Homo sapiens*, our own species. The study of the origin of mankind is called paleoanthropology (*paleo* means "ancient" and *anthropos* means "man"). In this chapter, we shall trace the erratic history of research in paleoanthropology.

The basic classification unit of animals and plants is the *species*. Groups of similar species form a *genus* (plural: *genera*), and similar genera form a *family*. Man-like species belong to the hominin family (formerly: "hominid" family). Bipedal locomotion is their distinguishing feature. Hominins walk upright on two legs. By examining the knee joints and pelvic structure of a fossil, scientists can generally determine whether a particular prehistoric creature walked on two legs.

## Errors of the Past

Unfortunately, studies of the evolutionary development of human beings have been plagued by profound misunderstanding. Until the 1960s, paleoanthropologists believed in the "single-species hypothesis." According to this hypothesis, the most primitive hominin species gradually evolved into a somewhat more advanced species, which in turn gradually evolved into

121

a still more advanced species, and so on. This process was repeated several times until it resulted in the most advanced hominin species of all: *Homo sapiens*, Modern Man. As the fossil of each new hominin species was discovered, it was interpreted as representing an additional phase in the straight-line evolutionary development from the earliest ape-like species to Modern Man.

The single-species hypothesis was finally discarded on the basis of extensive fossil evidence. In his book, *The Myths of Human Evolution*, Niles Eldredge, curator at the American Museum of Natural History, describes this hypothesis as follows[1]:

> *"the great evolutionary myth of slow, gradual, and progressive change. So bewitched were the single-species people by the linear elegance of this myth that they were reluctant to see it marred ... the standard expectation of human evolution — slow, steady, gradual improvement through time — is indeed a myth."*

## Genus *Australopithecus*

The picture of the hominin family that was generally accepted in the 1980s is now known to be erroneous. The hominin family was then thought to consist of only two genera, extinct *Australopithecus* and present-day *Homo*. The older genus, *Australopithecus* ("southern ape-man"), was thought to consist of only four species: *Australopithecus afarensis*, *A. africanus*, *A. boisei*, and *A. robustus*. The ancestral relationship between the various species of australopithecines was the subject of vigorous debate, with almost as many theories as paleoanthropologists.

During the 1990s, anthropologists identified three additional australopithecine species and, even more dramatically, they discovered a third hominin genus, *Ardipithecus*.[2] The importance of these findings does not lie in doubling the number of australopithecine species that once existed. The real importance of these findings lies in the fact that they totally changed our understanding of the history of prehistoric man.

In 1994, Yves Coppens, of the College of France in Paris, published an article entitled "The Origins of Humankind."[3] Coppens was a member of the French Academy of Sciences and a recognized authority on human evolution. In his article, Coppens presented a new theory for the evolutionary history of mankind. His basic idea was that the geographic isolation of Africa had triggered the evolution of hominins and that *"the Rift Valley in Africa holds the secret to the divergence of hominins from the great apes and to the emergence of human beings."*

The theory did not last long. Only three years later, Coppens's explanation was demolished by new fossil evidence. In 1997, two authorities on paleoanthropology (Alan Walker, of the American Academy of Sciences, and Meave Leakey, Head of Paleontology at the National Museum of Kenya) published an article explaining that Coppens's theory had been "debunked" by the newly discovered *bahrelghazali* hominin fossils.[4] These new fossils showed that there was *no* geographic isolation and that African hominins had lived on *both sides* of the African Rift Valley.

So it goes in the study of human evolution. Today's new theory for the origins of mankind is discarded on the basis of tomorrow's fossil discoveries.

David Pilbeam, of Harvard University, laments the large number of dramatic upheavals in the scientific understanding of human evolution that occurred within a very short time. After describing a complete change in scientific consensus that took place within a span of only five years, Pilbeam asks, "Why has the hominin fossil record been so badly misinterpreted?" He answers,[5] *"The early hominins were markedly different from any living species. In many instances, however, these differences were ignored and early hominins have been made to seem too much like modern humans."*

## Genus *Homo*

We now turn to the contemporary hominin genus, *Homo*. Since *Homo* fossils are more recent and more plentiful than the australopithecines, one would expect the scientific picture to be less confused. Unfortunately, that is not the case. The same scientific uncertainty and confusion that characterize the older *Australopithecus* fossils are also present regarding the more recent *Homo* fossils.

The genus *Homo* was once thought to consist of only three species, *Homo habilis*, *H. erectus*, and *H. sapiens* ("wise man"), with the last species being subdivided into two subspecies: Neandertal Man and contemporary Modern Man, the only living hominin.

Within a single decade, this picture totally changed. Genus *Homo* is now believed to consist of *seven* species.[6] Fossil discoveries have added *H. ergaster* and *H. rudolfensis*, and reclassification has added *H. heidelbergensis* and *H. neanderthalensis*. The lack of understanding regarding the early *Homo* fossils was the subject of an article appearing in the respected British journal *Nature*.[7] The article, which carries a satirical title — "Who is the 'Real' *Homo habilis*?" — describes the confusion surrounding the discovery of the hominin fossil known as OH-62. The article concludes, "*The new fossil rudely exposes how little we know about the early evolution of Homo.*"

## *Homo sapiens*

The most interesting scientific questions relate to *Homo sapiens* — our own species. Because *Homo sapiens* are the contemporary hominin species, one might expect the scientific evidence to be reasonably clear. In fact, strident arguments regarding the origins of our species are quite common among paleoanthropologists.

There are two opposing theories, each claiming to have found the correct explanation for the evolutionary history of contemporary human beings. These two competing schools of paleoanthropology use completely different methods of research. One school emphasizes traditional studies of prehistoric fossils. The opposing school uses the more modern method of studying DNA sequences to determine evolutionary history. It was hoped that the two complementary approaches to the study of human origins would lead to similar results, with each method confirming the findings of the other. But this has not been the case. The opposing schools have been at loggerheads for years.

One theory, designated "Out of Africa," claims that all human beings alive today descend from a single African woman ancestor, who lived about 200,000 years ago.

The opposing theory, designated "Multiregional," claims that the various human races arose independently at different times and at different sites around the world.

*Scientific American* invited each side to present its case in consecutive articles.[8,9] These articles make quite instructive reading. Each pair of respected authors claims that *their theory* has now been established *with absolute certainty.*

"Out of Africa" scientists wrote the following: *"After years of disagreement, we won the argument. The paleontologists admit that we had been right and they had been wrong."*

"Multiregional" scientists countered as follows: *"The fossil record is the real evidence for human evolution ... [We] describe a theory that synthesizes everything known about modern human fossils, archaeology, and genes."*

These two groups of scientists examined the *same data* and yet arrived at diametrically opposite conclusions! Such occurrences are common in the study of human origins. David Pilbeam, of Harvard University, laments,

*"Theories of human origins are relatively unconstrained by the fossil data."*

Another interesting feature of the *Scientific American* debate on the origins of *Homo sapiens* is that each side claims that the methods used by the opposing camp are *erroneous* and are based on *incorrect scientific procedures.*

"Multiregional" scientists write as follows of their opponents' methods of analysis: *"Their evidence was based on a flawed 'molecular clock' ... their hypothesis must be rejected because their reasoning is flawed."*

"Out of Africa" scientists describe their opponents' fossil data as follows: *"Fossils cannot be interpreted objectively ... The paleontologist's perspective contains a built-in bias that limits the power of observation ... The fossil record is infamously spotty ... and may lead down an evolutionary blind alley."*

How is one to relate to these completely opposing claims by well-known scientists? Is the genetic evidence *"inherently flawed"* or is it *"convincing"*? Are fossil remains and artifacts *"a monumental body of much more reliable evidence"* or do they *"contain a built-in bias"*? Are the genetic data *"complete and objective"* or do they *"rely on a long list of assumptions"*?

Perhaps one can assess the reliability of the opposing claims by considering the opinions of scientists who are not involved in the debate and, therefore, are presumably more objective in their views.

Evolutionary biologist Alan Templeton, of the University of Washington, is bitterly critical of the genetic and DNA data used by "Out of Africa" scientists[11]:

> *"It is likely that this ["Out of Africa"] hypothesis would never even have been proposed if a proper analysis had been performed on the original data set."*

Regarding the fossil analysis of "Multiregional" scientists, Erik Trinkaus, of the University of New Mexico, is very critical of their heavy reliance on fossil data. He points to the tendency of different scientists to look at the *same* fossils and come to *opposite* conclusions, with each scientist seeing the data as supporting his own theory[12]:

> *"The Saint-Cesaire [Neandertal] fossil was a perfect mirror, reflecting back into each viewer's eyes the convictions that he brought to it ... they constructed their hypotheses so flexibly that no evidence could possibly disprove them."*

## Conclusion

Perhaps the correct conclusion to be drawn from these claims and counterclaims is that *neither* side has a reliable theory. One can understand the comment of Trinkaus that paleontologists tend to see in the fossils whatever they wish to see[13]:

> *"What is uncanny — and disheartening — is the way in which each side can muster the same fossil record into an utterly different synthesis for human evolution. Reading their papers side by side gives the reader a distinct feeling of having awakened in a Kafka novel."*

# References

1. N. Eldredge and I. Tattersall, 1982, *The Myths of Human Evolution* (Columbia University Press: New York), pp. 2, 120.
2. T. D. White *et al.*, 1995, *Nature*, vol. 375, p. 88.
3. Y. Coppens, July 1994, *Scientific American*, pp. 62–69.
4. M. Leakey and A. Walker, June 1997, *Scientific American*, pp. 60–65.
5. D. Pilbeam, March 1984, *Scientific American*, pp. 63–69.
6. B. Wood, 1987, *Nature*, vol. 327, pp. 187–188.
7. I. Tattersall, January 2000, *Scientific American*, pp. 38–44.
8. A. C. Wilson and R. L. Cahn, April 1992, *Scientific American*, pp. 22–27. The article is entitled, "The Recent African Genesis of Humans."
9. A. G. Thorne and M. H. Wolpoff, April 1992, *Scientific American*, pp. 28–33. The article is entitled, "The Multiregional Evolution of Humans."
10. D. Pilbeam, 1980, in *Major Trends in Evolution*, editor L. K. Konigson (Pergamon Press: London), p. 267.
11. A. R. Templeton, 1992, quoted in *Scientific American*, May 1992, p. 80.
12. E. Trinkaus and P. Shipman, 1993, *The Neanderthals* (Jonathan Cape: London), p. 379.
13. E. Trinkaus and P. Shipman, 1993, *The Neanderthals* (Jonathan Cape: London), p. 390.

# Chapter 20

# Are Contemporary Human Beings Continuing to Evolve?

## Introduction

The answer to the question posed in the title of the chapter would *seem* to be in the affirmative. Hominins — the man-like species — first appeared about five million years ago, and they have been evolving continuously ever since. Genus *Australopithecus* evolved into the early species of the current genus *Homo*. The early *Homo* species then evolved into our own species, *Homo sapiens* or Modern Man. We here discuss the question of whether *Homo sapiens* will continue to evolve into yet another species.

*It is our thesis that contemporary human beings will not undergo any further evolution at any time in the future.*

## The Evolution of New Species

In order to understand why evolution for contemporary human beings has ceased, one must first review what is meant by the term "evolution." Every creature is defined by its array of genes. Evolution occurs and a new species appears when the array of genes undergoes such a significant change that the "new" creature can no longer interbreed with the previous creature to produce fertile offspring. In other words, *evolution implies a significant change in the array of genes.*

In the process of reproduction, the array of genes is copied. When a gene is copied, an error sometimes occurs, and one speaks of a genetic mutation. Most genetic mutations cause a *defect* in the creature, leading to its inability to function and to produce healthy offspring. Therefore, such a genetic mutation will disappear from the gene pool. Sometimes, however, the genetic mutation *enhances the ability of the creature to produce healthy offspring*. Such a mutation is termed a "favorable" mutation. Favorable mutations tend to be incorporated into the species gene pool. After many generations, as many favorable genes are incorporated into the species gene pool, a new species will be formed.

The key point is that evolution requires favorable mutations. A favorable mutation may make the individual creature stronger, faster, smarter, less susceptible to disease, etc. This mutation therefore enhances the creature's chances of survival.

## The Struggle for Existence

One of the basic principles of evolution is the "*struggle for existence.*" The term refers, for example, to the available food supply. For any given species, the number of offspring far exceeds the available food supply. Not all individuals in a species have the same physical ability to obtain the scarce food. Some individuals are stronger or faster than others because their array of genes contains the "favorable" genes that make them stronger or faster. This leads to a "*struggle.*" Individuals that possess such "favorable genes" have a better chance of obtaining the scarce food and thus surviving. This is called "*survival of the fittest.*"

But what if every individual in the species had the same ability to obtain food, regardless of whether the individual was strong, fast, and smart or weak, slow, and dull? Then, there would be no "*struggle for existence.*" In that case, there would be no favorable genes because there would be no advantage in being faster or stronger. The array of genes in the species would never change and there would be no evolution.

*As will now be explained, this is the situation for contemporary human beings.*

# The Neolithic Revolution

About ten thousand years ago, *a revolution occurred in human behavior.* This sudden, dramatic event is known as the Neolithic Revolution (see Chapter 21 for details). This Revolution marked the *sudden appearance of civilization.*

Before the Neolithic Revolution, the only available food for human beings was the animals and the plants that they could hunt or gather. Archaeologists refer to this stage as the "hunter-gatherer" stage. During this stage, food was always in short supply. There was always a struggle to obtain the needed food.

But as a result of the Neolithic Revolution, humans learned to greatly increase their food supply by applying modern methods of agriculture and animal husbandry. Today, human beings plant vast fields of grains, fruits, and vegetables. They raise large herds of cattle and other animals. Moreover, human beings have learned how to preserve their food to make it available throughout the year. Nowadays, *human beings determine the size of the available food supply.* There is now no difficulty in producing as much food as desired. For the first time in history, human beings are capable of controlling the food supply of the planet.

The *"struggle for existence"* is now over. Contemporary human beings do not need to "struggle" to obtain their food. With a better distribution system, there would be enough food for everyone. There are no longer any "favorable" genes enhancing one's ability to obtain food. With a few exceptions, everyone is equally *"fit"* to obtain his food. The evolutionary principle of *"survival of the fittest"* no longer applies.

But there still appears to be a problem. The nineteenth-century British economist Thomas Malthus claimed that increased populations will eventually outstrip the food supply, no matter how much food is produced. This problem has now been solved. Contemporary human beings have the ability to control their population through family planning. China's "one-child" policy illustrates how efficiently and quickly a government program can reduce the size of the population.

*It is within the power of contemporary human beings to ensure that there will always be an adequate food supply for everyone and that hunger will cease to exist.* To implement this program, human beings do not need better genes through evolution. They must provide a more efficient distribution system and a more humane society.

The *"struggle for existence"* also relates to the ability of an individual to defend himself against predators. In the past, being fast and strong was important in the struggle against predators. Today, effective defensive weapons are readily available.

The *"struggle for existence"* also relates to an individual being less susceptible to disease. Here, too, the problem has to a large extent been solved. "Favorable" genes are not the key to avoiding disease. Modern medicine can solve most health problems.

## Genes That Improve Intellectual Capabilities

Thus far, we have discussed "favorable" genes that improve the physical capabilities of humans, such as being stronger, faster, or less susceptible to disease. What about genes that improve one's intellectual capabilities, termed "smartness" genes? The past ten thousand years have witnessed enormous technological progress. It would seem that genes that lead to this technological progress are "favorable" in the sense that they enhance one's ability to survive. Doesn't a person with "smartness" genes have an advantage in the "struggle for existence" over those individuals without such genes? Doesn't that imply that human evolution will continue as "smartness" genes propagate throughout the population?

Modern society has made the enormous medical and technological progress of recent years *available to everyone.* One does not have to be a "smart" engineer to obtain a refrigerator to store one's food or be a "smart" surgeon to benefit from life-saving operations. "Smartness" genes are not necessary to survive.

An essential feature of all the technological progress during the past ten thousand years is the unique ability of contemporary human beings to communicate abstract ideas to each other. This enables everyone to benefit from the accomplishments of their predecessors.

The importance of communication can hardly be overestimated. The many technological innovations that have revolutionized human society resulted from the cumulative efforts of many talented and dedicated people. Because contemporary human beings are capable of communicating abstract ideas to each other, a scientist need not "reinvent the wheel" before making new contributions. The ability to continue to build on the work of others has led to the rapid technological progress that is the hallmark of civilization. Everyone benefits from the "smartness" genes of the few.

For example, Isaac Newton is universally recognized as the greatest physicist of all time. Nevertheless, my physics students know more physics than Newton ever did. Not because they are smarter than Newton or because they possess "smartness" genes that Newton lacked. The reason is that they have learned Newtonian physics *in addition to* the physics that has been developed by many Nobel Prize winners in the centuries since Newton.

Newton himself once remarked, "*If I have seen further than others, it is by standing on the shoulders of giants.*"

## Summary

Since the advent of civilization about 10,000 years ago, contemporary human beings have been able to survive without being physically strong, fast, or smart. Therefore, changes in one's array of genes that might make one stronger, faster, or smarter are not necessary to enhance one's chances for survival. It therefore follows that for contemporary human beings, *a significant evolutionary change in one's array of genes is unlikely to occur in the future.*

# Chapter 21

# Are Contemporary Human Beings Unique Creatures? The Neolithic Revolution

## Human Uniqueness

Contemporary human beings are the only creatures who write books, compose music, tell jokes, pass laws, discuss philosophy, formulate mathematical theorems, invent complex instruments, hold national elections, and the list goes on and on. We are unique.

Moreover, contemporary human beings dominate all other vertebrate species, including fishes, amphibians, reptiles, birds, and mammals. In fact, contemporary human beings dominate so effectively that many countries have passed laws to prevent humans from wiping out these other species. Our destructive power seems almost unlimited — cutting down vast forests, causing global warming, polluting the atmosphere, and destroying the ozone layer. We are unique both in our creative and in our destructive powers.

However, the idea that we humans are unique creatures has been vigorously denied by some scientists who insist that human beings are just another one of the two million species of animals thus far identified. These scientists admit that we are different from other species, but claim that every species possesses some special properties that set it aside as a separate species. It is only human pride — so they claim — that makes us think that we are unique creatures.

Scientist and historian Jared Diamond, of the University of California at Los Angeles (UCLA), has written a book, *The Third Chimpanzee*, whose theme is the lack of uniqueness in human beings. The title refers to human beings, whereas the other two species are the common chimpanzee and the bonobo chimpanzee. Diamond asserts that human beings have no special talents that are not shared to some extent by many other animals. Even our ability to think does not fundamentally distinguish us from the other species of chimpanzees. We are more talented, of course, but not unusually so.

The uniqueness of human beings is easily demonstrated. A Google search reveals that UCLA, Diamond's home university, has a student enrollment of about 40,000 and the libraries contain about eight million books. However, *not a single one of the 40,000 UCLA students is a chimpanzee! And not a single one of the eight million books in the UCLA libraries was written by a chimpanzee!* Ironically, Diamond's own book is the best evidence that human beings are not just another species of chimpanzee. No chimpanzee has ever written a book comparing human beings to chimpanzees!

The above-mentioned facts seem surprising in view of the close physical similarity between the two species. Diamond points out that 98.5% of the genetic material of humans is also found in chimpanzees. Since genes determine the physical characteristics of an animal, this close genetic similarity demonstrates that we are very similar to the chimpanzee *physically*. This raises the following question. If we are so very similar to the chimpanzee *physically*, why are we so very different *intellectually and culturally*?

This question is not restricted to chimpanzees. It applies equally well to prehistoric man. No prehistoric man ever founded a university, wrote a book, established a government, or displayed any other feature of civilization during the five million years of the existence of prehistoric man.

The same question also applies to Modern Man (*Homo sapiens*, our own species). Modern Man first appeared about 200,000 years ago. However, the earliest signs of civilization appeared only a few thousand years ago. What took so long?

# Neandertal Man and Modern Man

The hominins that are most similar to Modern Man are the Neandertals, who were the prehistoric people who immediately preceded Modern Man. They lived for over 200,000 years throughout Europe, Western Asia, and the Middle East. Then, for unknown reasons, the Neandertals suddenly disappear from the fossil record. *Only* Modern Man is found in archaeological sites that date from the last 30,000 years. (Other names given to peoples who lived during this period, such as Cro-Magnon Man, refer to *cultural* groups, and not to *physical* types of people. Cro-Magnon Men were as much Modern Men as contemporary Frenchmen or the Chinese.)

Scientists agree that Modern Man did *not* evolve from the Neandertals. Ian Tattersall, of the Department of Anthropology at the American Museum of Natural History, is a recognized authority on Neandertal Man[1]:

*"Homo heidelbergensis quite likely gave rise to the Neandertals, whereas a less specialized population founded the lineage that produced Modern Man."*

Erik Trinkaus, of the University of New Mexico, shares this view[2]:

*"Modern humans did not evolve out of the local Neandertal population."*

Our principal sources of information regarding the culture of prehistoric men are the tools and other artifacts found in their ancient campsites. A comparison between prehistoric tools and those of Modern Man would not be meaningful, because the large well-developed brain of Modern Man gives him an obvious advantage over small-brained earlier hominins. But there is one exception to the general pattern of earlier hominins having smaller brains. That exception is Neandertal Man. Trinkaus writes,[3]

*"Neandertals had brains as large and as complex as our own."*

Were the Neandertals physically deficient or limited in any way that might have hindered their cultural development? Trinkaus writes,[4]

> *"Neandertals were physically not less human than Modern Men. They had the same postural abilities, manual dexterity, range and character of movement as did Modern Men ... a much stronger grip than Modern Men, but there was nothing gorilla-like about it. Their control of movement was the same as ours."*

The foregoing discussion shows that it is indeed meaningful to compare the cultural achievements of Modern Man with those of Neandertal Man. Significant cultural differences between the Neandertals and ourselves cannot be explained in terms of the physical limitations of Neandertal Man.

## Neandertal Culture

What were the accomplishments of Neandertal Man? What great cities did he build? What profound writings did he leave for posterity? What important moral teachings did he expound? What marvelous paintings, stirring musical compositions, magnificent sculpture, moving poetry, breathtaking architecture, beautiful gardens, and profound scientific discoveries remain from the Neandertals to mark their 200,000-year-long sojourn on our planet? *Their meager cultural legacy contains none of these items!*

Neandertal tools were primarily flints with a sharp edge. Their tools look quite similar to the sharp stones that one finds strewn along every beach. Tattersall explains as follows[5]:

> *"The stone-working skills of the Neandertals consisted [of using] a stone core, shaped in such a way that a single blow would detach a finished implement. They rarely made tools from other materials. Archaeologists also question the sophistication of their hunting skills. Despite misleading earlier accounts, no clear evidence has ever been found for symbolic behavior among Neandertals or for the production of symbolic objects. Even the occasional Neandertal practice of burying their dead may have been simply to discourage hyena incursions, or*

*have a similar mundane explanation, for Neandertal burials lack the 'grave goods' that would attest to ritual and belief in an afterlife ... **The Neandertals lacked the spark of creativity that distinguishes Modern Man.***"

Regarding artistic accomplishments, it is important to point out that the magnificent cave paintings found in southwestern France, Spain, and elsewhere were *all* the work of early Modern Man. No cave painting produced by a Neandertal has ever been discovered.

What are the reasons for Neandertal Man's lack of culture? Why was Modern Man able to revolutionize all aspects of his environment, while Neandertal Man hardly left a trace of his existence? In fact, archaeologists must search hard to find *any* remnants of Neandertal Man. Recall that the Neandertal brain "*does not suggest any differences [from Modern Man] in intellectual or behavioral capabilities.*" Scientists have no convincing explanation for the great disparity in culture and intellectual capabilities between these two hominin species that were so similar physically.

## The Culture of Modern Man

In discussing the impressive culture that characterizes Modern Man, we need not limit ourselves to the latest technological developments. As soon as he appeared, Modern Man demonstrated his *enormous* cultural superiority over Neandertal Man. Scientific accounts emphasize the various far-reaching technological advances *immediately* introduced by Modern Man[6]:

*"The toolmaking industries of Modern Man are completely different from those of Neandertal Man ... reflecting a quantum leap in mental abilities ... **Modern Men who followed the Neandertals were their intellectual superiors in every way.***"

Recall that the average brain size and complexity were the same for the two species. Regarding physical capabilities, the Neandertals were equals of modern humans. The primitiveness of Neandertal culture is one of the mysteries surrounding these physically quite advanced but culturally backward predecessors of Modern Man.

## The Neolithic Revolution

Modern Man gradually developed his technological and artistic skills. Then, about 10,000 years ago, an explosion of cultural innovations occurred, covering all aspects of human behavior. The cumulative effect of all these changes was to revolutionize human society. Archaeologists refer to the sudden appearance of so many technological and artistic advances as the *Neolithic Revolution* or the *Agricultural Revolution*.[a]

The Neolithic Revolution was so all-encompassing that it has become *the major milestone* in prehistoric chronology. Archaeologists denote all earlier times as Paleolithic (Old Stone Age), whereas subsequent times are denoted as Neolithic (New Stone Age).

The many fundamental cultural innovations that occurred during or shortly after the Neolithic Revolution include agriculture, animal husbandry, metalworking, the wheel, the first written language, ceramic pottery, weaving, prepared foods (bread, wine, cheese, and butter), musical instruments, and advanced architecture, to name but a few. This vast proliferation of cultural advances permitted the formation of the complex social organization that soon gave rise to the first cities and thus to modern civilization (the term means "city-making"). The enormous range of these profound cultural and social developments is emphasized in every archaeological account of this period:

> *"A crucial event in human history was the beginning of agriculture 10,000 years ago in the Near East. The accumulation of surplus food supplies enabled large settlements to be established, resulting in the emergence of Western civilization."*[7]

---

[a] Strictly speaking, there are some differences between the Agricultural Revolution and the Neolithic Revolution, and these two did not occur at exactly the same time in each locality. However, such subtleties are of more interest to the professional archaeologist than to the layman, and we shall here treat these two revolutionary events as equivalent.

*"Agriculture and animal husbandry appeared at roughly the same time ... Technological progress, the mastery of new materials (such as metals) and new energy sources (such as wind and water power) ... The acceleration of human history cannot be better illustrated than by comparing the changes of the past 10,000 years with those of the previous four million years."*[8]

*"One cannot avoid being impressed at how rapidly the transition occurred from paleolithic hunting groups to regionally organized communities ... domestication of plants and animals, establishment of farming communities, development of pottery ... Bronze tools and weapons were produced. Writing evolved from pictographic notations ... specialized artisans made quantities of diverse goods ... market centers became towns ... The urban revolution was underway, the world of people was radically transformed, and the first civilizations were taking shape."*[9]

*"The development of plant and animal domestication is referred to as the Neolithic Revolution ... The changes arising from food production so altered human life that all manner of new developments came into being ... village ways of life, population growth, and complex forms of social organization."*[10]

*"Major characteristics of agricultural economies were evident ... advanced cultivation techniques occurred with explosive consequences ... populations increased enormously ... the pace of change was so rapid and so far-reaching."*[11]

What were the causes of the Neolithic/Agricultural Revolution? What triggered all these *"explosive," "far-reaching,"* and *"revolutionary"* changes that so altered human society? The fact is that no one really knows[12]:

*"What factors caused a shift away from hunting-gathering to food production? The question continues to be debated by archaeologists and anthropologists."*

# The Uniqueness of Mankind

The most striking features of human uniqueness occurred in the following areas: intellectual curiosity, communication, and morality.

## *Intellectual Curiosity*

Man is the only species that displays intellectual curiosity regarding matters that do not enhance his chances for survival. All other species concern themselves solely with food, shelter, safety, mating, and so on, for themselves and their family, tribe, or colony. By contrast, human beings express great interest in and devote much time to the pursuit of knowledge and art that have no practical consequences.

An excellent illustration of this phenomenon is the book that you are now holding in your hands. Reading this book will *not* increase your salary, will *not* put better food on your table, and will *not* in any way improve your physical situation. Nevertheless, in spite of the complete absence of any tangible benefit, you continue to read in order to satisfy your intellectual curiosity.

## *Communication*

The past few thousand years have witnessed the enormous progress made by human beings in all areas of intellectual endeavor. An essential feature of this progress is the unique ability of human beings to communicate abstract ideas to each other. This enables them to benefit from the accomplishments of their predecessors.

The importance of communication can hardly be overestimated. The many technological innovations that have revolutionized human society resulted from the cumulative efforts of many talented people. Because human beings can communicate abstract ideas, a scientist need not "reinvent the wheel" before making new contributions. The ability to build on the work of others has led to the rapid technological progress that is the hallmark of civilization.

## *Morality*

The most striking feature of man's uniqueness lies in the realm of morality. Man's morality extends far beyond his own species. Human beings are concerned about the welfare of other species and they have established extensive programs worldwide to help other species, especially those designated as endangered.

Man's most important expression of morality concerns his own species. Man is capable of making decisions based on abstract principles of right and wrong. Humans may sacrifice their personal welfare, and indeed even their lives, in the cause of morality.

The plight of starving people in Africa generates a worldwide appeal for help. Most Africans have nothing at all in common with the average American or European — neither race nor religion nor language nor ideology nor lifestyle. Yet, the sight of starving children on the TV screen touches the hearts of viewers, and their conscience "demands" that they contribute money to alleviate the suffering.

Of all the species in the animal kingdom, only man deals with moral problems. And only man possesses the faculty for making moral judgments. This privilege and the accompanying responsibility are ours alone.

# References

1. I. Tattersall, April 1997, *Scientific American*, p. 52.
2. E. Trinkaus and P. Shipman, 1993, *The Neandertals* (Jonathan Cape: London), p. 414.
3. E. Trinkaus and P. Shipman, 1993, *The Neandertals* (Jonathan Cape: London), p. 418.
4. E. Trinkaus and W. W. Howells, December 1979, *Scientific American*, p. 99.
5. I. Tattersall, January 2000, *Scientific American*, p. 43.
6. N. Eldredge and I. Tattersall, 1982, *The Myths of Human Evolution* (Columbia University Press: New York), pp. 154, 159.
7. S. Lev-Yadun *et al.*, June 2000, *Science*, vol. 288, p. 1602.
8. S. L. Washburn, September 1978, *Scientific American*, p. 154.
9. E. A. Hoebel and T. Weaver, 1979, *Anthropology and the Human Experience* (McGraw-Hill: New York), pp. 183, 195, 201.
10. G. H. Pelto and P. J. Pelto, 1979, *The Cultural Dimensions of the Human Adventure*(Macmillan: New York), p. 93.
11. A. Sherratt, editor, 1980, *The Cambridge Encyclopaedia of Archaeology* (Cambridge University Press: Cambridge), p. 407.
12. J. Diamond, 1997, *Guns, Germs, and Steel* (W. W. Norton: New York and London), p. 109.

# Part 3

# Blunders of Great Scientists

Even the greatest scientists occasionally made blunders in their scientific work. By the word "blunder," we refer to a gross error or a serious misunderstanding. The following chapters will describe some famous blunders in the fields of physics and evolutionary biology.

# Chapter 22

# Isaac Newton (1642–1727)

## Isaac Newton

Isaac Newton is widely considered to be the greatest physicist of all time. He established the laws of mechanics (called Newtonian mechanics in his honor), the law of gravity, the principles of optics, the principles of color, calculus (the standard method of performing mathematical calculations to this day), and much more. He wrote the most important book of physics ever, entitled *Principia* (full English title: *Mathematical Principles of Natural Philosophy*, where "natural philosophy" is the old name for "science"). Newton explained planetary motion, one of the most important scientific problems of his time. Johannes Kepler had formulated the empirical laws of planetary motion in 1609, but before Newton, no one could *explain* these laws.

It would take an entire volume to fully describe Newton's fundamental work in physics and mathematics. Indeed, it is difficult to exaggerate the importance of Newton's contributions to science. Newton lived to age 85, never married, and worked hard all his life to develop his many original ideas. His work was central to the Scientific Revolution of the seventeenth century, the intellectual transformation that produced modern science.

Newton was also a deeply religious person and was very interested in theology. He studied various theological writings and published a commentary on the Book of Daniel. It was his great interest in theology that led to his scientific blunder.

## Proof for the Existence of God: The Prime Mover Argument

Seeking proof for the existence of God may sound quaint to the modern ear, but it was a matter of utmost importance to medieval theologians and religious scientists. The most widely quoted proof for the existence of God is known as "the prime mover argument" (the word "argument" does not denote a dispute. Rather, "argument" is an old English word for "proof"). It is instructive to analyze this proof and discuss its shortcomings.

Experience seems to teach us that no object moves without a cause for its motion. Aristotle asserted, "There is no motion without a mover." Examples of this simple truism abound. When I rearrange the furniture, I am aware of the fact that the couch will not move even one millimeter unless I continue to push. No matter how energetically one throws a ball, the ball eventually comes to rest. The reason that the ball continues to move even after it leaves one's hand was thought to be due to the "impetus" that one imparts to the ball. But when the ball uses up its "impetus," it comes to rest because "There is no motion without a mover."

The sun, moon, stars, and planets have always been observed to be in motion. It follows, therefore, that some agency ("mover") must be moving the heavenly bodies along their paths. This agency, called "the prime mover," can only be God, because no other power is able to move the heavenly bodies around the sky. Thus, the *observed* perpetual motion of the heavenly bodies *proves* the existence of God.

## Refutation

Among its many topics, Newton's *Principia* contains the laws of motion. Newton's first law of motion — the law of inertia — states that, in complete contrast to Aristotle, a moving object *will continue to move forever* unless some force causes the object to *stop moving*. In the examples given earlier, the force that causes the furniture or the ball to stop moving is the force of friction. However, if there were no friction present, then their motion would persist *forever*. In the heavens, there is no friction. Therefore, according to Newton's law of inertia, the heavenly bodies *will continue to move forever without any agency being required to keep them moving*. The prime mover argument for the existence of God *is thus refuted*.

Newton's law of inertia predicts straight-line motion, whereas the planets move around the sun in an ellipse. This was explained by Newton as being due to the gravitational attraction between the sun and the planets. The elliptical orbits of the planets follow from Newton's laws, *without the need to invoke divine intervention*.

When Newton realized that his scientific findings had refuted the primary proof for the existence of God, Newton was troubled. After giving the matter much thought, Newton "solved" the problem in the following ingenious way.

## The Blunder

The path of a planet around the sun is an ellipse because of the gravitational attraction of the sun. However, this is true if the planet experiences the gravitational attraction of *only the sun*. But each planet is also affected by the gravitational attraction of all the other planets. Because the sun is a thousand times more massive than all the planets combined, this additional attraction is very weak. Nevertheless, it may have a significant effect on planetary motion. To test this matter, Newton set out to calculate the effect on planetary motion due to the gravitational attraction of the other planets. This is an extremely difficult problem, known as the many-body problem, which cannot be solved exactly. Therefore, Newton introduced an approximation that he considered adequate and calculated the effect on planetary motion. The result was startling! Newton found that the solar system was *unstable*. The small gravitational attraction exerted by each planet on every other planet disrupted the stability of the entire solar system. According to Newton's calculations, each planet will slowly drift away from the other planets and from the sun. Thus, the sun will eventually be devoid of planets and the solar system will cease to exist.

But thousands of years of stargazing had established that the solar system is, in fact, stable. The planets do not drift apart over the course of time. How could the contradiction between Newton's calculations and the observations be resolved?

The explanation given by Newton was that it must be God who maintains the stability of the solar system! Newton proposed that divine intervention pushes the planets back when they start to drift away from the sun.

A hundred years later, the brilliant French mathematician Pierre-Simon de Laplace reexamined the problem of the stability of the solar system. Laplace showed that Newton's approximation was inadequate. Laplace introduced a more accurate approximation, solved the resulting equations, and found that the solar system *was stable* after all. In other words, the instability of the solar system claimed by Newton did not actually exist. It was only the result of Newton's inadequate approximation.

# Chapter 23

# Albert Einstein (1879–1955)

## Albert Einstein

Albert Einstein is widely considered to be the second-greatest physicist in history, after Isaac Newton. Einstein had a *miracle year* in 1905, in which he published *four* of the most important papers of the twentieth century. These include his theory of special relativity, his explanation of the photoelectric effect which confirmed quantum theory, his vital contribution to atomic theory, and his famous formula $E = Mc^2$, which established the relationship between matter and energy. He received the Nobel Prize for explaining the photoelectric effect. It is difficult to understand why Einstein did not receive a second Nobel Prize for his special theory of relativity and even a third Nobel Prize for his theory of general relativity, which is actually a new theory of gravity.

Einstein continued to make major contributions to physics. In 1915, Einstein published his theory of general relativity, which has been described by the Russian physicist Lev Landau as *"the most beautiful theory in all of physics."* Einstein's theory of general relativity is a theory of gravity and also a theory of the structure of the universe.

## The Blunder

Einstein was one of the early pioneers of quantum theory. However, when the theory was developed further, Einstein did not accept its strange ramifications. Quantum theory is *probabilistic*. A *probabilistic* theory means that quantum theory is unable to predict *with certainty* the results of an experiment. One can perform the exact same experiment twice and yet *obtain different results each time.*

Einstein could not accept the idea that the laws of physics do not lead to definite predictions. Einstein expressed his rejection of the probabilistic nature of quantum theory by saying, *"God does not play dice."* It is now known that the probabilistic nature of quantum theory is correct.

The equations of Einstein's theory of general relativity describe the structure of the universe. According to Einstein's equations, the universe must always be expanding or contracting. The universe can never remain static. This was a very disturbing prediction because all scientists, including Einstein, believed that the universe was, in fact, static and unchanging.

In the past, Einstein had always accepted the results of the equations that he developed. But this time, he deviated from his usual practice. A universe that was perpetually expanding or contracting was just too unbelievable for Einstein to accept.

Einstein decided that his equations should be altered. The suggested alteration was to add a term, which is called the *cosmological term*. When the cosmological term is added to Einstein's equations, there is a solution that corresponds to a static universe. Thus, thought Einstein, the problem had been corrected.

In 1929, on the basis of a detailed study of the galaxies, the astronomer Edwin Hubble, of the Mount Wilson Observatory, announced that the universe was expanding. He had discovered that all galaxies are moving away from each other.

When Einstein learned of Hubble's discovery of the expanding universe, he realized that his original equations had been correct all along and had not needed any additional term. The universe was expanding, just as his *original* equations had predicted. He referred to his adding the cosmological term to his original equations as *"the biggest blunder of my professional career."*

# Chapter 24

# Charles Darwin (1809–1882)

## Charles Darwin

There is hardly a person who has not heard of Charles Darwin's theory of evolution. This is one of the most famous theories in the entire realm of science. Darwin's book, *The Origin of Species*, was published in six editions, has been reprinted innumerable times, and has been translated into most of the world's languages.

Darwin's scientific contribution was *not* that he proposed that present-day animals were the descendants of previous animals that had become extinct. This had long been known from the fossils that were discovered everywhere. Darwin's contribution was to propose a mechanism to explain *how* previous animals had evolved into today's animals.

Darwinian evolution occurs through a favorable chance mutation in the genetic array of an animal. This appearance of a favorable mutation enhances the animal's chances for survival by making the animal a bit stronger, faster, smarter, less susceptible to disease, etc. An individual animal with a favorable mutation has a greater chance than other individual animals to live long enough to reproduce the next generation. In this way, the favorable mutation will become incorporated into the gene pool of the species. The accumulation of many favorable mutations over many generations eventually brings about significant changes in the animal, leading to an entirely new species.

The key point of Darwin's theory of evolution is that evolution works in small cumulative steps, through vast periods of time to cause a primitive species to evolve very gradually into a more complex species.

## The Problem

Darwin was convinced that if his theory is correct, then the intermediate forms between the ancient primitive species and the contemporary more complex species must appear in the fossil record. However, Darwin was aware that the predicted intermediate forms had never been found.

Darwin devoted an entire chapter of his book to this problem. Darwin faced this problem openly, writing, "*Why, then, is not every geological formation and every stratum full of intermediate links … This, perhaps, is the most obvious and serious objection that can be urged against the theory.*"

Darwin answered as follows: "*The explanation lies, as I believe, in the extreme imperfection of the geological record.*" (now called the "fossil record.") Darwin was convinced that the required intermediate links did exist "*in every geological formation and in every stratum*" and they would be found if one were to make a thorough search.

However, during the last century, careful searches have been carried out and it has become quite clear that every geological formation is *not full of intermediate links*. Intermediate links have occasionally been found, such as the archaeopteryx, an intermediate link between dinosaurs and birds. But discovering a few intermediate links does not fulfill Darwin's requirement that *"every geological formation"* should be *"full of intermediate links."*

Darwin considered this requirement to be so important that he wrote the following: *"He who rejects this view of the imperfection of the geological record, will rightly reject the entire theory."*

## The Blunder

Darwin had linked his theory of evolution to gradualism. We now know that gradualism rarely occurred in the evolution of the higher taxonomic levels of order, class, or phylum. Instead, there is extensive fossil evidence of mass extinctions, the Cambrian explosion of lifeforms, and punctuated equilibrium (see Chapters 15 and 16), all of which are inconsistent with gradualism.

This does not mean, however, that one should *"rightly reject the entire theory,"* as Darwin had written. Darwin's theory contains many important features of the mechanism of evolution that are valid even in the absence of gradualism. With certain modifications, Darwin's theory admirably accounts for the fossil record.

# Chapter 25

# Marie Curie (1867–1934)

## Marie Curie

Marie Curie was one of the greatest scientists of her generation. She was awarded two Nobel Prizes, the first in Physics in 1903 and the second in Chemistry in 1911. Only three other scientists have been awarded two Nobel Prizes in science: John Bardeen, Frederick Sanger, and Karl Sharpless. Marie Curie also received the Davy Medal, Matteucci Medal,

Actonian Prize, Albert Medal, Willard Gibbs Award, and many other prizes and awards. Element 96 is named "curium" in her honor, and the basic unit of radioactivity is the "curie." Marie Curie pursued her scientific work with relentless dedication, often working under extremely difficult conditions.

Marie Curie suffered all her life due to severe prejudice against women. Even after she won the Nobel Prize, she was not offered a professorship at the University of Paris. Not only was Marie a woman but she was also Polish, another severe defect in the eyes of the French scientific establishment. (Curie was her married name. Her maiden name was Sklodowska.) Even after she received her second Nobel Prize in 1911, the French Academy of Sciences still failed to elect Marie Curie to membership. When invited by the Royal Institution in London to give a lecture on her Nobel-Prize-winning research on radioactivity, she was not allowed to speak because she was a woman and her husband Pierre gave the lecture. (Marie and Pierre had shared the 1903 Nobel Prize for their joint research on the newly discovered phenomenon of radioactivity.) The Nobel Prize Committee had at first intended to award the 1903 Nobel Prize *only* to her husband because Marie was a woman. However, one member of the Committee was so outraged by this intention that Marie's name was included in the award.

Marie was awarded a second Nobel Prize in 1911 for discovering two new chemical elements, polonium and radium. She was forced to perform the very difficult chemical research outside in the open, in winter and summer, because she was not allocated a laboratory by the University of Paris.

## The Blunder

The history of her discovery of these two elements is as follows. It was known that the element uranium is radioactive. This meant that uranium nuclei eject particles. At first, no one knew what the different ejected particles were, and so they were named alpha, beta, and gamma particles. (It was later discovered that alpha particles are helium nuclei, beta particles are electrons, and gamma particles are photons.)

Marie Curie studied pitchblende, which is an ore that is slightly radioactive because it contains small amounts of the radioactive element uranium. Marie Curie was convinced that the measured radioactivity of pitchblende ore was too strong to be explained solely by the uranium it contained. She thought that there must be an additional radioactive element in the pitchblende ore. Therefore, she obtained a large amount of pitchblende ore and began the difficult task of purifying it in order to isolate the other suspected radioactive element. For various technical reasons, this was an enormously difficult task that required years of patient work. Finally, Marie had purified the pitchblende sufficiently to confirm the existence of a second radioactive element in addition to uranium. She sent her evidence to the appropriate international committee and they approved her evidence. As its discoverer, Marie was given the privilege of naming the new element. As an ardent Polish nationalist, Marie named the new element "polonium," in honor of her beloved homeland Poland.

During the following weeks, Marie continued her chemical work of purifying the pitchblende ore. She then realized that there was yet another radioactive element in the pitchblende ore, which exhibited far stronger radioactivity than polonium. Once again, she sent her evidence to the appropriate international committee, who approved it. And once again, she was given the privilege of naming the second radioactive element. But she had already used the name polonium for the first element that she had discovered. Therefore, she named the second element "radium," because of its strong radioactivity.

Thus, polonium became the name of the less important radioactive element, a name generally unknown to the public. Radium became the name of the much more important radioactive element, a name familiar to every educated person. Although Marie Curie had nerves of steel while conducting her research, she did not have the patience to wait a few more weeks to complete her chemical purification before announcing her discovery. Thus, she missed the opportunity of honoring her beloved homeland Poland with the name of one of the most well-known elements in the period table of chemical elements.

# Chapter 26

# Alan Turing (1912–1954)

## Alan Turing

Alan Turing was a mathematician, computer scientist, cryptologist, and theoretical biologist. He is known primarily for his pioneering work in computers and is widely considered to be the father of computer science and artificial intelligence. He provided a formalism for the concepts of algorism and computation with his Turing machine, which was the first model of a general-purpose computer. He designed the first stored-program computer, and he proved that the "halting problem" for Turing machines is undecidable. Turing demonstrated that there are some mathematical yes–no questions that can never be answered by computation. Turing developed the idea of a "Universal Turing Machine" that could perform any task that is computable. His many accomplishments in the field of computer science include "Turing's proof," "the Turing pattern," and "Turing reduction."

Alan Turing received many awards and also features on the Bank of England £50 note.

The most famous of his contributions are in the field of cryptology. Turing broke the code of the German enigma cipher machine during World War II. Intercepting the German coded messages and decoding them enabled the Allies to defeat the Axis powers in many crucial engagements, including the Battle of the Atlantic. To this end, Turing developed a new decryption machine.

In order to decide whether or not a particular computer is thinking, Turing devised what has become known as the "Turing Test." The test is as follows: If the user of a computer cannot discern whether he is communicating with a computer or with a person, and he is, in fact, communicating with a computer, then that computer is thinking.

## The Blunder

John Searle, of the University of California at Berkeley, showed that Turing had made a fundamental error regarding the Turing Test. Searle's thesis is that no computer could ever think because a computer only *manipulates symbols*, whereas thinking involves attributing *meaning to symbols*. For example, when a computer defeated the world chess champion, the computer did not "know" what chess was, it did not "know" that it had won the chess match, and it did not "know" that it had become the new chess champion. In other words, the winning computer was incapable of giving any meaning to its victory in the chess match. The computer had simply manipulated the symbols that led to its victory. *This is not what is generally meant by thinking.*

To demonstrate his thesis, Searle devised an ingenious counter-example, known as the "Chinese Room Argument." However, the explanation of the "Chinese Room Argument" and its refutation of the Turing Test lie beyond the scope of this book.

# Chapter 27

# Lev Landau (1908–1968)

## Lev Landau

Lev Landau was one of Russia's greatest physicists. He made fundamental contributions to the quantum theory of diamagnetism, superfluidity, superconductivity, plasma physics, electrodynamics, neutrinos, second-order phase transitions, and more. There is hardly an area of physics in which he did not publish important research. In 1962, he was awarded the

Nobel Prize. Explaining the importance of Landau's many contributions is far beyond the scope of this book.

In addition, Landau was the author of a series of textbooks covering all aspects of theoretical physics. His books have been translated into English and are widely used in universities throughout the world. He also founded a school of theoretical physics in Russia. His students include many of the leading Russian theoretical physicists of the twentieth century.

## Liquid Helium and Landau

The blunder of Landau concerns his research on liquid helium. In order to explain this blunder, we must first describe the properties of liquid helium.

At room temperature, the element helium is a gas, similar to other gases, such as oxygen and nitrogen. As the temperature is lowered, every gas eventually becomes a liquid. As the temperature is lowered even further, the liquid eventually becomes a solid. But this is not the case for helium. Unlike oxygen and nitrogen, *helium never becomes a solid and remains a liquid even at the lowest possible temperature (absolute zero), which is –273°.*

This is not the only strange property of liquid helium. Another weird property of liquid helium is known as "superfluidity." This term means that liquid helium can flow through even the tiniest hole, as though it has no friction. *No other liquid behaves in this manner.* In addition, liquid helium exhibits a host of other peculiar properties. All these properties are due to the quantum behavior of liquid helium. The challenge was to correctly apply the principles of quantum theory in order to explain the weird behavior of liquid helium.

## Liquid Helium and the Bose Transformation

Studies of the quantum properties of gases showed that, in principle, at a temperature close to absolute zero, a gas will undergo a strange transformation. This is called a "Bose transformation" or a "Bose condensation" (named after the Indian physicist Satyendra Nath Bose). The Bose transformation causes the gas to exhibit many unusual properties.

However, a Bose transformation is never expected to occur in nature because, at these extremely low temperatures, no gas exists in nature. Every gas has already become a solid or, in the case of helium, a liquid.

It was soon realized that all the weird properties that were observed for *liquid helium* exactly corresponded to the weird properties predicted *for a gas* that undergoes a Bose transformation. This convinced scientists to propose that one is able to explain the bizarre properties of liquid helium if one ignores the fact that liquid helium is a liquid and not a gas. One should pretend, so to speak, that helium remains a gas and never becomes a liquid at any temperature. Upon making this assumption, *the Bose transformation applies to helium, and all the bizarre properties of liquid helium are explained.*

## The Blunder

Landau refused to accept this idea. He insisted that since low-temperature helium was a *liquid*, it is absurd to pretend that helium *remains a gas at all temperatures*. Therefore, according to Landau, the Bose transformation could not apply to liquid helium and could not be used to explain the bizarre properties of low-temperature liquid helium.

But this was a blunder. It took a long time for Landau to finally concede that the strange properties of liquid helium were indeed due to a Bose transformation. This was the case even though the Bose transformation does not normally apply to a liquid. However, liquid helium is an exception.

# Chapter 28

# Niels Bohr (1885–1962)

## Niels Bohr

Niels Bohr was a dominant figure in the early development of quantum theory. He made fundamental contributions to the understanding of atomic structure. He also explained why radiation is emitted by energetic atoms. Bohr combined Rutherford's proposed structure of the atomic nucleus with Planck's quantum theory of light in order to develop what has become known as the Bohr model of the atom. For this important research, he was awarded the 1922 Nobel Prize. The numerous other scientific awards that Bohr received include the Copley Medal, Hughes Medal, Matteucci Medal, Franklin Medal, and the Max Planck Medal.

Bohr's name is also associated with many important concepts in quantum theory, including the Bohr radius, Bohr magneton, Bohr–Sommerfeld theory, Bohr–Van Leeuwen theorem, and Bohr–Kramers–Slater theory. Bohr became the elder statesman of quantum theory. Many leading physicists, including Nobel Prize winners, traveled to Copenhagen to carry out research with him at his Institute of Theoretical Physics at the University of Copenhagen.

Bohr proposed the principle of complementarity, according to which experiments may be analyzed in terms of contradictory principles. The most famous example was wave–particle duality. This example relates to the fact that electrons, as well as light, sometimes seem to behave like particles and sometimes seem to behave like waves. This contradiction was considered by Bohr to be a basic aspect of quantum theory and was a central feature of what is known as the Copenhagen interpretation of quantum theory, in honor of the location of Bohr's research center.

## The Blunder

Nobel laureate Richard Feynman conclusively demonstrated that Bohr's concept of wave–particle duality does not exist. Feynman refers to the concept of wave–particle duality as "a state of confusion."

Since electrons have a definite mass and a definite charge, it is clear that electrons are particles. Niels Bohr's claim that electrons also exhibit wave-like properties appears to be supported by the two-slit experiment. In this experiment, electrons show an "interference pattern." It was incorrectly believed that an interference pattern can only be produced by waves, but not by particles. This statement is true for classical physics, but *classical physics is incorrect. All particles*, including electrons, follow the laws of quantum theory. According to quantum theory, *particles can indeed produce an interference pattern* (see Chapter 4). In fact, *there is no experimental evidence that supports the idea that electrons are a wave phenomenon.* Bohr's Copenhagen interpretation, which claims that electrons sometimes exhibit wave-like properties, is a blunder.

This is also true for light. *There is no experimental evidence that supports the idea that light is a wave phenomenon.* Light always behaves like a stream of particles.

Unfortunately, due to Bohr's great influence, wave–particle duality is sometimes still mentioned in scientific articles. This incorrect concept of wave–particle duality is described in Google as one of Bohr's important contributions to quantum theory. Even Google makes mistakes. The blunder lives on.

# Chapter 29

# Lord Kelvin (William Thomson) (1824–1907)

## Lord Kelvin (William Thomson)

Lord Kelvin was one of Britain's foremost scientists — a physicist, mathematician, and engineer. He was a professor at the University of Glasgow for 53 years. The University of Glasgow has a permanent exhibition of the works of Kelvin, including many of his original papers, instruments, and other artifacts. He was awarded the prestigious Copley Medal of the Royal Society of London and served as its president from 1890 to 1895. William Thomson was the first British scientist to be elevated to the House of Lords. He took the title Lord Kelvin after the Kelvin River which flows near his laboratory at the University of Glasgow.

Kelvin determined the correct value of absolute zero temperature as $-273.15°$, and the units of the absolute temperature scale are named "Kelvins" in his honor. The Joule–Thomson effect is also named in his honor. Kelvin carried out important mathematical analyses of electricity and formulated the first and second laws of thermodynamics. He contributed significantly to unifying physics, which was then in its infancy.

Kelvin also had a career as an engineer and inventor, which propelled him into the public eye and earned him wealth, fame, and honors. For his work on the transatlantic telegraph cable, Kelvin was knighted in 1866 by Queen Victoria. In recognition of his achievements, William Thomson was ennobled in 1892 to become Lord Kelvin.

## The Age of the Earth

One of the most controversial questions in science in the nineteenth century was the age of the Earth. We now know that the correct age is 4.55 billion years. However, in 1863, Kelvin published a calculation that implied that the Earth was only 50 million years old. This figure was far too low to be acceptable to geologists or evolutionary biologists. However, they could not find any flaw in Kelvin's calculation.

## The Blunder

Kelvin assumed, *correctly*, that the Earth began as a very hot sphere with a molten center. But he also assumed, *incorrectly*, that the Earth was continually cooling, with no added source of heat. With this assumption and some known data, Kelvin could calculate when the cooling began. Accordingly, this would be the age of the Earth.

Kelvin did not realize that *there was a source of heat inside the Earth,* continually providing additional heat. After Henri Becquerel discovered radioactivity, scientists realized that the heat provided by the decay of radioactive elements in the interior of the Earth must be included in Kelvin's calculation. When one includes the heat produced by the radioactive elements, one obtains the modern value of 4.55 billion years as the age of the Earth.

Kelvin also made another blunder. In his 1900 address to the British Association for the Advancement of Science, Kelvin declared the following:

> *"There is nothing new to be discovered in physics now. All that remains is to make more precise measurements."*

Kelvin's statement was made shortly before some of the greatest discoveries in the history of physics, including quantum theory, relativity theory, chaos theory, string theory, black holes, dark matter, and dark energy. We do not know what exciting new discoveries await mankind.

# Chapter 30

# Enrico Fermi (1901–1954)

## Enrico Fermi

Enrico Fermi was the greatest Italian physicist of the twentieth century. Fermi developed the statistics required to explain a large class of subatomic particles. This type of statistics is now known as Fermi statistics and the relevant subatomic particles are now known as "fermions." The United States National Accelerator Laboratory was named "Fermilab" and element 100 was named "fermium" in his honor. He was one of the chief architects of the nuclear age. Fermi made important progress in clarifying the nuclear reactions caused by neutrons. He also performed important research in general relativity and statistical mechanics, and he developed the theory of beta decay. Fermi was the charismatic leader of an important group of physicists who worked together in Italy.

Research in physics is carried out in two different modes: theoretical physics and experimental physics. The theorist explains physical phenomena by working with pencil and paper and with the computer. The experimentalist teases out the secrets of nature by making measurements. Fermi was a theorist. He was the first professor of theoretical physics at the newly established prestigious Accademia d'Italia.

Later in his career, Fermi also carried out experimental research. He discovered that it is possible to produce a new element by bombarding a nucleus with slow neutrons. If the neutrons are fast, they simply bounce off the nucleus that they hit. But if one slows down the neutrons, Fermi found that the nucleus absorbs the slow neutrons and thereby produces a new element. Fermi announced that, in this way, he had produced elements 93 and 94, which are beyond the period table of elements that appear in nature, which ends with uranium, element 92. For this important work, Fermi received the 1938 Nobel Prize.

After traveling to Stockholm to receive his Nobel Prize, Fermi continued to the United States. He never returned to Italy because Italy had passed anti-Semitic laws similar to the anti-Semitic laws of Nazi Germany. Since Fermi's wife Laura was Jewish, he feared for her safety. In the United States, Fermi joined the Physics Department of Columbia University and later the University of Chicago. At the University of Chicago, Fermi produced the first controlled nuclear chain reaction in December 1942.

## The Blunder

It was subsequently discovered that Fermi had not actually produced elements 93 and 94, as he had announced. The nuclei that Fermi bombarded with slow neutrons *had split into lighter elements*. Fermi had erroneously thought that the bombarded nuclei that had absorbed the slow neutrons *became transformed into heavier elements.*

It should be emphasized that no one doubts that Enrico Fermi deserved a Nobel Prize for all his important research accomplishments. But there is a certain irony in the fact that the Nobel Prize Committee chose to award the Nobel Prize to Fermi on the basis of the one piece of his research that he had misinterpreted.

# Chapter 31

# Robert Boyle (1627–1691)

## Robert Boyle

Robert Boyle was one of the greatest scientists of the seventeenth century. He was both a physicist and a chemist. Boyle was a Fellow of the Royal Society. He is regarded as the first modern chemist and one of the pioneers of the modern experimental scientific method. He is best known for Boyle's law, which established the relationship between the pressure, volume, and temperature of a gas. His book, *The Skeptical Chymist*, was an important fundamental work in the field of chemistry.

## Oxygen and the Corrosion of Metals

It is common experience that metals corrode (in iron, such corrosion is called rusting). The correct explanation for corrosion was given in 1789 by the French chemist Antoine Lavoisier, the discoverer of the element oxygen. Lavoisier demonstrated that corrosion occurs when oxygen reacts with a metal to form a metallic oxide, which often appears powdery. If the corroded metal is heated in the presence of carbon, the oxygen of the corroded metallic oxide combines with the carbon to form carbon dioxide gas which escapes into the atmosphere, restoring the pure metal.

## Phlogiston and the Corrosion of Metals

In Boyle's lifetime, a hundred years before Lavoisier, the accepted theory of corrosion was totally different. The incorrect but then-accepted theory of corrosion was based on a substance called "phlogiston." The idea was as follows:

Metals have a shiny, attractive appearance because they are rich in phlogiston. But as the metal ages, the phlogiston escapes from the metal into the atmosphere. The lack of phlogiston causes the metal to become powdery and to lose its attractive metallic appearance. However, the shiny metallic luster can be restored by heating the metal with carbon, which is very rich in phlogiston. Upon heating with carbon, the corroded metal absorbs phlogiston from the carbon and regains its shiny metallic appearance.

This theory seemed very compelling because it explained the major features of corrosion. However, the phlogiston theory of corrosion was completely wrong. There is no such substance called "phlogiston."

In 1630, when the phlogiston theory was still widely accepted, John Rey decided to weigh phlogiston. He took a piece of metal and carefully weighed it. He then allowed the metal to corrode. Its subsequent loss of weight would correspond to the weight of the phlogiston that had escaped from the metal during corrosion. However, to his great surprise, Rey found that the metal had *gained weight* upon corrosion. How could the *loss* of phlogiston cause an *increase* in weight?

## The Blunder

Boyle thought of an ingenious explanation for this puzzle. He proposed that phlogiston has *negative weight*! Therefore, *losing* phlogiston would result in an *increase in weight*.

This "explanation" may seem absurd to the modern ear, but many "scientific" ideas that were accepted in the seventeenth century now seem absurd.

The correct explanation for the increase in weight is that corrosion is caused by the *addition* of oxygen to the pure metal. Upon the addition of oxygen, a metallic oxide is formed. This *addition* of oxygen causes the *observed increase in the weight* of the corroded metal.

# Chapter 32

# Ernest Rutherford (1871–1937)

## Ernest Rutherford

Ernest Rutherford, a pioneer researcher in both atomic and nuclear physics, has been described as "the father of nuclear physics" and as the "greatest experimentalist since Michael Faraday." He was awarded the 1908 Nobel Prize. Other prizes that he received include the Rumford Medal, Barnard Medal, Cresson Medal, Copley Prize,

Franklin Medal, Matteucci Medal, and Faraday Medal. Element 104 was named "rutherfordium" and there is a certain unit of energy that is called a "rutherford."

Rutherford's numerous scientific achievements include the concept of radioactive half-life, the discovery of the atomic nucleus, artificially induced nuclear reactions, and the discovery and interpretation of what has become known as Rutherford scattering.

Rutherford is most famous for his scattering experiment. It was known that the atom consisted of negative charges (now known to be electrons) and an equal quantity of positive charge. However, it was unclear how this positive charge is distributed within the atom. According to the then-accepted incorrect Thomson model, the positive charge was thought to be distributed uniformly throughout the atom.

Rutherford thought of an ingenious experiment to test the Thomson model. This is now known as a scattering experiment. Rutherford took an extremely thin gold foil, only a few atoms thick. (Gold can be hammered into a thinner foil than any other metal.) Rutherford bombarded the gold foil with the positively charged alpha particles (now known to be helium nuclei) that are spontaneously emitted from some radioactive elements. Most alpha particles passed right through the gold foil. But sometimes an alpha particle would be scattered by a gold atom. Rutherford measured the *angle* through which the charged alpha particle was scattered by the positive charge in the atom. This enabled him to determine how the positive charge is distributed in the atom.

Rutherford was amazed to find that an alpha particle was occasionally scattered through a very large angle. This could only happen if all the positive charge in the atom was concentrated within a very small volume, now known as the "nucleus." This model of the atom was very different from the Thomson model.

If all the positive charge of the atom is concentrated in the nucleus, then, thought Rutherford, the negatively charged electrons are probably continuously circling the nucleus. This model of an atom could be compared to a miniature solar system, with the nucleus playing the role of the sun and the electrons playing the role of the planets that circle the sun. The "Rutherford model of the atom" was thus born.

# The Blunder

It was soon recognized that there was a problem with the Rutherford model of the atom. If a charged particle — the electron — moves in a circle, it will radiate energy and fall into the nucleus, causing the atom to collapse. However, atoms do not collapse and their electrons do not fall into the nucleus. Rutherford ignored this problem and assumed that, for unknown reasons, there exist mysterious "stationary electron orbits" in which the electrons do not fall into the nucleus and that they do not radiate energy.

Rutherford's model of the atom — in which electrons circle the central nucleus — became widely accepted. It wasn't until quantum theory was developed that the true behavior of electrons in atoms was finally understood. It then became clear that *electrons do not circle the nucleus.*

Unfortunately, the incorrect Rutherford model of the atom — with the electrons circling the central nucleus — is still quite popular and appears widely.

Another blunder of Rutherford's concerned the possibility of obtaining a useful source of energy by means of splitting the atom. Rutherford insisted that splitting atoms could not possibly be a source of energy, saying, in his unique way, "Anyone who looks for a source of energy in the splitting of atoms is *talking moonshine.*" Today, hundreds of nuclear power stations around the world generate electrical energy precisely by splitting atoms.

# Chapter 33

# Henry Fairfield Osborn (1857–1935)

## Henry Fairfield Osborn

Henry Fairfield Osborn was a leading American paleontologist (researcher of fossils). He served as the Director of the American Museum of Natural History in New York for over 25 years. Under his leadership, this museum became world famous. Osborn led many fossil-hunting expeditions and was honored by having two fossils named after him, a dinosaur and an

African dwarf crocodile. He was the senior vertebrate paleontologist for the United States Geological Survey. Osborn accumulated one of the finest fossil collections in the world and he was universally recognized as "a great paleontologist."

Osborn received many awards, including the Hayden Memorial Geological Award, Cullum Geographical Medal, Wollaston Medal, and Daniel Giraud Elliot Medal from the American National Academy of Sciences.

## *Hesperopithecus*

There was one fossil that had a profound effect on Osborn's career. This fossil bears the scientific designation *hesperopithecus* ("western ape-man") to emphasize that this was the first hominin fossil ever discovered in the Western Hemisphere (near Snake Creek in the State of Nebraska). It will be the rare reader who has heard of *hesperopithecus*. The history of this fossil has been shoved deep under the rug by paleontologists, and its story has been carefully excised from all scientific writings — with good reason. Of all the blunders that were committed by evolutionary biologists during the twentieth century in their studies of hominin fossils, none can compare with that of *hesperopithecus*!

Our story begins in 1922, when a geologist discovered an ancient-looking tooth. The geologist sent the tooth to Osburn, the expert in fossils, to solicit his opinion regarding the importance of the tooth. Osborn's enthusiasm warmed as he studied the tooth and considered its implications. Osburn created a sensation by claiming that this tooth must have come from a direct human ancestor, the first hominin fossil found in America.

Osborn named the fossil *hesperopithecus* ("western ape-man") and announced it to the scientific world in a paper published in the April 1922 issue of the prestigious journal *Proceedings of the National Academy of Sciences*.

In order to emphasize the importance of *hesperopithecus*, Osborn commissioned a graphic reconstruction of a *hesperopithecus* couple in a forest of Snake Creek fauna. The impressive reconstruction was given a place of honor in the American Museum of Natural History. This was a marvelous example of the lifelike three-dimensional exhibits for which this museum is justly famous. Looking at a photograph of the *hespero-pithecus* exhibit, one is amazed by the many details of the physical appearance and the cultural behavior of this prehistoric man and woman that Osborn claimed to have deduced *from one single tooth.*

## The Blunder

Five years later, Osborn's world collapsed. Additional fossil evidence discovered later near Snake Creek in Nebraska showed without doubt that the *hesperopithecus* fossil was, in fact, the tooth of a *pig.*

Among the unfortunate results of this scientific fiasco was the fact that over several years, a million visitors to the American Museum of National History in New York had been enthralled by the brilliantly exe-cuted reconstruction of the *hesperopithecus* prehistoric man and woman who had supposedly lived in the Nebraskan forest. These visitors would never read the scientific literature that revealed the truth about *"the man who was actually a pig."*

# Chapter 34

# Marcellin Boule (1861–1942)

## Marcellin Boule

Marcellin Boule was the leading expert in human paleontology (study of fossils) in France in the early decades of the twentieth century. Boule was Director of the Institute for Human Paleontology at the famous Museum National d'Histoire Naturelle of France. He was also the editor of the major journal *L'Anthropologie* and the founder of two other scientific journals.

## Neandertal Man

Boule's special area of research was Neandertal Man, the prehistoric human who preceded Modern Man (*Homo sapiens*). ("Neandertal," which means Neander Valley, derives from the valley near Dusseldorf, Germany, where the first fossil skull was discovered in 1856.) Neandertal fossils have since been discovered throughout Europe, the Middle East, and even further afield, with hundreds of nearly complete skeletons now available for study.

The Neandertals first appeared about 300,000 years ago. For unknown reasons, they disappeared from the fossil record about 30,000 years ago. Since that time, only Modern Man appears in the hominin fossil record.

## Boule's Analysis of Neandertal Man

In 1910, Boule presented his preliminary analysis of a complete Neandertal fossil. Boule then published a definitive monograph on the Neandertals in the 1911–1913 issues of *Annales de Paleontologie*. The monograph immediately became a classic, a study of such thoroughness that it established the paleontology of humans as a scientific discipline.

Boule characterized the Neandertals as brutish, bent-kneed, and not fully erect bipeds. In an illustration that Boule commissioned, the Neandertal was characterized as a wild, hairy, gorilla-like figure. Because of Boule's description, "Neandertal" became a term of abuse. The conduct of hooligans is often described as "no better than Neandertals," thus conveying an image of uncouth and uncivilized conduct. These assertions are confirmed in *Roget's Thesaurus*. The list of synonyms for "Neandertal" includes terms such as "savage," "brutal," "bestial," and "animal." The image of Neandertals is that of coarse-featured, stooped-walking, long-armed brutes, whom no one would want to meet in a dark alley.

# The Blunder

Unfortunately, Boule's classic monograph on Neandertal Man *was incorrect in every respect.* In his book on hominin paleontology, appropriately entitled *The Myths of Human Evolution*, Niles Eldredge, curator of the American Museum of Natural History in New York, writes the following:

> *"Every feature that Boule stressed in his analysis had no basis in fact ... To Boule, the premier French paleontologist of his day, we owe the shambling brutish image of the Neandertals, immortalized in a thousand comic strips."*

Erik Trinkaus, of the University of New Mexico, also emphasizes this point:

> *"What is remarkable is that Boule's monograph is astonishingly wrong in many of its conclusions ... Boule reconstructed the vertebral column of the Neandertals as much less straight than it was, giving rise to a stooping posture and slouching gait, forwardly thrust head and perpetually bent knees. The drawing in his monograph imprinted itself on the minds of anthropologists everywhere. It was the perfect troglodyte, the brute, the savage."*

Scientists now understand that Neandertal Man probably looked remarkably similar to Modern Man. It is said that if a Neandertal were to board a train, his fellow passengers would not even be aware that sitting next to them was a member of a different species!

How could a famous scientist such as Marcellin Boule make such gross errors? The explanation lies in the lack of objectivity with which Boule approached his study. He began his research with preconceived notions of what the Neandertals *should look like.* Boule believed that

Neandertals had nothing to do with human ancestry, and his analysis succeeded in expelling these brutish forms from the human family tree.

As Trinkaus explains,

> *"Boule did not deliberately and knowingly slant his results. But he only
> saw what was agreeable, and was oblivious to elements that suggested
> otherwise."*

Worst of all was the effect of Boule's erroneous work on his colleagues. Eldredge notes that *"Boule's authority was so close to absolute
that his conclusions strongly affected paleontological thinking for several
decades."* Trinkaus adds that *"Boule's erroneous conclusions had a more
lasting effect on the image of Neandertals than any previous work."*

# Chapter 35

# Sir Grafton Elliot Smith (1871–1937)

# Sir Arthur Keith (1864–1944)

# Sir Arthur Smith Woodward
# (1866–1955)

# Sir Grafton Elliot Smith, Sir Arthur Keith, and Sir Arthur Smith Woodward

Sir Grafton Elliot Smith was a Fellow of the Royal Society. Smith received many awards, including the Royal Medal of Britain, French Legion of Honor, Struthers Prize, Imperial Ottoman Order of the Osmaniah, Gold Medal of the Royal College of Surgeons, and Huxley Memorial Medal of the Royal Anthropological Institute. He was a professor of anatomy at University College, London.

Sir Arthur Keith was also a Fellow of the Royal Society. Keith was President of the Royal Anthropological Institute, editor of the *Journal of Anatomy*, conservator of the Hunterian Museum of the Royal College of Surgeons, and International Member of the American Academy of Sciences.

Sir Arthur Smith Woodward was also a Fellow of the Royal Society. Woodward was President of the Geological Society and Secretary of the Paleontographical Society. His awards include the British Royal Medal, Lyell Medal, Clarke Medal, Wollaston Medal, Hayden Geological Award, Linnean Medal, and Thompson Medal of the American Academy of Arts and Sciences.

These three scientists were variously characterized as "the great names of the British school of paleontology of the 1920s and 1930s" and "the three leading lights of British anthropology and paleontology." Each man was a recognized world authority. Smith and Keith were the foremost British anatomists of their day and Woodward was an expert in hominin paleontology. Each man was knighted by his monarch as a sign of the esteem accorded him by the scientific community.

# Piltdown Man

Piltdown Man was the most spectacular fraud of twentieth-century science. In 1908, an amateur fossil collector named Charles Dawson announced that, in a Piltdown gravel pit on the Sussex coast of England, he had found parts of the fossil head of a prehistoric man that soon became known as Piltdown Man. Before his sudden death in 1916,

Dawson "discovered" a few more pieces of Piltdown Man's skull and jaw. The Piltdown fossils were accepted as genuine by the scientific community and were formally given the Latin name *Eoanthropus dawsoni* ("Dawson's Dawn Man") in honor of its discoverer.

Piltdown Man was a fake! Dawson had combined a modern human skull with the jaw of a modern ape (an orangutan), both of which he had stained to match the color of the Piltdown gravel pit. Dawson broke off those parts of the skull where it attaches to the jaw to hide the fact, otherwise obvious, that the (human) skull did not fit the (ape) jaw. He filed down the ape's teeth a bit to make them look more human, and in various ways contrived to make the bones look ancient, as befits a prehistoric fossil.

What is important is *not* the fact that a hoax had been perpetrated. Every profession has its cheats. What is central to our discussion is the assessment of these fraudulent fossils by the leading members of the scientific community. One would have thought that as soon as this jaw of an ape reached the hands of the professional anatomists, the game would be up. How could any skilled anatomist fail to immediately recognize that the Piltdown jaw was identical in every respect to the jaw of a modern orangutan and that the Piltdown skull was identical in every respect to the skull of a contemporary human? How could the experts not realize that Piltdown Man had none of the features that characterize "prehistoric man"? Surely, Dawson's fraud would be exposed by leading anatomists within a matter of minutes.

But that did not happen. In fact, this fraud remained undetected *for forty years*! Piltdown Man created a sensation and it was universally accepted as genuine. From 1912 until 1953, every scientific reference book informed its readers of the great importance of Piltdown Man in evolutionary history. We were told that Piltdown Man was our direct ancestor.

# The Blunder

The most important supporters of the Piltdown fraud were Sir Grafton Elliot Smith, Sir Arthur Keith, and Sir Arthur Smith Woodward. Each of these leading scientists was convinced that the Piltdown fossils were genuine! The combined influence of Woodward, Keith, and Smith ensured that Piltdown Man became accepted by the scientific community. Over the course of nearly half a century, over two hundred scientific articles were published discussing the important role of Piltdown Man in the history of human evolution.

It was obvious to anyone who examined the Piltdown fossils that the skull (which had belonged to a modern man) appeared much more human-like and much less ape-like than the jaw (which was the jaw of a modern ape). To explain this anomaly, scientists invoked the concept of "mosaic evolution," asserting that different parts of the body evolve at different rates.

The British school of paleontology insisted that the skull of Modern Man evolved relatively rapidly, whereas the jaw evolved more slowly. Thus, according to the British school, it was expected that our prehistoric ancestors would, at some early stage, have a relatively modern human-like skull while still sporting a relatively primitive ape-like jaw. When Piltdown Man seemed, in the eyes of the British paleontologists, to display precisely these characteristics, he was warmly welcomed.

But the basically human skull of Piltdown Man should also display some ape-like features, and the basically ape-like jaw of Piltdown Man should also display some human features. After all, Piltdown Man was *supposed* to be a fossil in transition — on the way to becoming a Modern Man. *These completely non-existent features were exactly what each of the three leading British scientists claimed to see in the Piltdown fossils!*

For example, according to world-famous anatomist Sir Grafton Elliot Smith,

> *"The Piltdown skull, when properly reconstructed, is found to possess strongly simian [ape-like] peculiarities. These features harmonize completely with the simian features of the jaw, which have been exaggerated by most writers ... The outstanding interest of the Piltdown skull lies its confirmation of the view that in the evolution of man, the brain led the way."*

In other words, Britain's leading anatomist claimed to see distinctly human anatomical features in the jaw of a modern orangutan and distinctly ape-like anatomical features in the skull of a contemporary human being. In fact, *none of these anatomical features existed.* He "saw" what he hoped to see.

## Chapter 36

# Blunders of the Scientific Community: Treatment of Women Scientists

One of the less complimentary features of the scientific community was the shabby treatment given to outstanding women scientists. Among the many unworthy incidents, we shall focus on four of the most blatant.

## Marie Curie (1867–1934)

Marie Curie was one of the greatest scientists of her generation. She was awarded two Nobel Prizes, the first in Physics in 1903 and the second in Chemistry in 1911. Only three other scientists have ever been awarded two Nobel Prizes in science.

Unfortunately, in spite of her outstanding scientific research, Marie Curie suffered for much of her life due to the severe prejudice against women displayed by her colleagues.

Although she won the 1903 Nobel Prize, she was not offered a professorship at the University of Paris. Not only was Marie a woman but she was also Polish, another severe defect in the eyes of her French scientific colleagues. (Curie was her married name. Her maiden name was Sklodowska.) Even after being awarded a second Nobel Prize in 1911, the French Academy of Sciences still failed to elect Marie Curie to membership. When invited by the Royal Institute in London to give a lecture on her Nobel Prize-winning research on radioactivity, she was not allowed to speak because she was a woman and her husband Pierre gave the lecture.

Some of her colleagues went so far as to write to the Nobel Prize Committee suggesting that they not include her in the 1903 Nobel Prize in Physics because Marie was a woman. However, one member of the Committee was so outraged by this suggestion that Marie's name was included in the award.

Marie was awarded a second Nobel Prize in 1911 for discovering two new chemical elements, polonium and radium. Because she was not allocated a laboratory by the University of Paris, she was forced to perform extremely difficult chemical research outside in the open, in winter and summer. Marie Curie pursued her scientific work with relentless dedication, but she was often forced to work under extremely difficult conditions.

Only later in life was Marie Curie given the scientific recognition that she so richly deserved. She eventually received the Davy Medal, Matteucci Medal, Actonian Prize, Albert Medal, Willard Gibbs Award, and other prizes and awards. Element 96 is named "curium" in her honor, and the basic unit of radioactivity is the "curie."

# Jocelyn Bell (1943–)

Jocelyn Bell made one of the most important discoveries in astronomy in recent times. She discovered the first pulsar. Pulsars are neutron stars that emit strong pulses of radiation with a very precise interval between the pulses. Measurements taken of a binary pulsar star system were used to confirm the existence of gravitational radiation, which was first predicted by Einstein's theory of general relativity. Certain types of pulsars serve as the most accurate clocks in the universe.

The history of the discovery of pulsars is as follows: Jocelyn Bell was an Irish doctoral student in the laboratory of Antony Hewish at the University of Cambridge.

The laboratory had a newly built radio telescope. Astronomers had always observed the stars with optical telescopes, that is, measuring the visible light that stars emit. However, Antony Hewish realized that stars also emit radio waves, and measuring these radio waves could give new and important information about the star. Therefore, his laboratory commissioned the construction of a telescope that measured radio waves, called a radio telescope. Bell helped design this radio telescope.

As part of her doctoral research, Bell analyzed data obtained from the radio telescope. On 6 August 1967, Bell noticed some strange signals picked up from the telescope. Her supervisor Hewish dismissed these signals as meaningless, caused simply by instrument noise. However, Bell was convinced that the signals that she measured were real and important and not just instrument noise. She persisted in her conviction and repeated the measurements again and again, night after night, until she finally convinced Hewish. Thus, Bell had discovered the first pulsar.

A few months later, on 21 December 1967, Bell discovered a second pulsar. When measurements from different radio telescopes confirmed her results, all doubts were finally laid to rest. It was now undeniable that Bell had indeed discovered a pulsar.

In 1974, the Nobel Prize in Physics was awarded to Antony Hewish "for pioneering research in radio astrophysics and for his decisive role in the discovery of pulsars." However, *the Nobel Prize Committee did not include Jocelyn Bell in the award, even though it was she, and not Hewish, who had discovered the first pulsar.* In fact, it had been difficult for Bell to convince Hewish that she had indeed discovered a pulsar. He kept insisting that the radio signals she measured were meaningless instrument noise.

Twenty years later, in 1993, the Nobel Prize in Physics was awarded to *both* Joseph Taylor *and* his graduate student Russell Hulse for their important confirmation of Einstein's theory of general relativity. In this case, however, the Nobel Prize Committee saw no barrier to awarding the Nobel Prize *to the graduate student together with his professor.* But of course, in 1974, the graduate student was female, not male.

Ultimately, Jocelyn Bell did receive the scientific recognition that she so richly deserved. Her many prizes and awards include the Special Breakthrough Prize in Fundamental Physics, Oppenheimer Prize, Faraday Prize, Herschel Medal, Copley Medal, and Gold Medal of the Royal Astronomical Society.

# Lise Meitner (1878–1968)

Lise Meitner was an outstanding physicist in an era when outstanding women physicists were a rarity. Like other women scientists, she experienced extensive prejudice because of her sex. In addition, living in Germany, she was subject to anti-Semitism because she was Jewish. The fact that Meitner had converted to Christianity had no influence on the Nazis when they stripped her of her research positions. Fearful for her personal safety, Meitner fled to Sweden.

The most blatant example of the prejudice against Lise Meitner concerned the 1944 Nobel Prize in Physics. The background is as follows: When Otto Hahn bombarded uranium atoms with neutrons, he obtained strange results that he did not understand. He sent his results to Meitner in Sweden, his long-time collaborator, seeking an explanation. Lise Meitner correctly understood that the results demonstrated that the uranium atoms had split into lighter atoms, a process now known as nuclear fission. The process of nuclear fission is, of course, central to nuclear reactors and nuclear energy.

In the years following the Nazi rise to power in Germany, Hahn had become an important, well-connected member of the scientific community, whereas Meitner had become a Jewish woman refugee, holding no academic position and having little influence in the scientific power structure. In addition, Hahn had implied to colleagues that Meitner had not contributed significantly to the discovery of nuclear fission. He described her work as that of a laboratory technician. As a result, Meitner was ignored in the award of the 1944 Nobel Prize in Physics, which was given only to Hahn.

According to the Nobel Prize archives, which were opened in the 1990s, Meitner was nominated for a Nobel Prize nearly fifty times over the years, but to no avail. Her nominees included many outstanding scientists who were Nobel laureates themselves.

In 1997, an article bearing the title "Nobel Tale of Injustice" was published in *Physics Today*, the journal of the American Physical Society. According to this article,

> *"Lise Meitner did not share the 1944 Nobel Prize because the structure of the Nobel committees was ill-suited to assess interdisciplinary work;*

*because the members of the Committee were unable or unwilling to judge her contribution fairly; and because during the war, the Swedish scientists relied on their own limited expertise. Meitner's exclusion from the award may well be summarized as a mixture of disciplinary bias, political obtuseness, ignorance, and haste."*

Ruth Sime published a detailed biography of Lise Meitner, entitled *From Exceptional Prominence to Prominent Exception.* Sime wrote the following:

*"In Sweden there was no sympathy for refugees from Nazi Germany. The country was small, with a weak economy and no immigrant tradition, and its academic culture had always been firmly pro-German, a tradition that did not change much until the middle of the war when it became obvious that Germany would not win. During the war, members of Siegbahn's group saw Meitner as an outsider, withdrawn and depressed. They did not understand the displacement and anxiety common to all refugees, or the trauma of losing friends and relatives to the Holocaust, or the exceptional isolation of a woman who had single-mindedly devoted her life to her work."*

In later life and after her death, Lise Meitner finally received the honor that she deserved. She was praised by Einstein as the "German Marie Curie." Meitner received the Max Planck Medal, Exner Medal, Austrian Decoration for Science, Pour le Merite, Foreign Member of the American Academy of Arts and Sciences, and Enrico Fermi Award. Element 109 was named "meiterium." Craters on the moon and Venus and schools and streets in Austria and Germany have been named in her honor.

# Rosalind Franklin (1920–1958)

The research work of Rosalind Franklin contributed significantly to establishing the double-helix molecular structure of DNA. Franklin's expertise lay in X-ray photography, and she provided truly outstanding X-ray photographs of crystalline DNA. In particular, her X-ray *Photo 51* provided vital information used by Francis Crick and James Watson in their research work that demonstrated the double-helix molecular structure of

DNA. The 1962 Nobel Prize in Physiology or Medicine was awarded to Crick and Watson for their success in this project. Unfortunately, Rosalind Franklin had already died of cancer in 1958 at the young age of 37. Since Nobel Prizes are not awarded posthumously, it was not possible to include Franklin in the Prize.

Understanding the double-helix structure of DNA played a vital role in laying the foundation for the science of genetics. Sections of the long DNA molecule form the genes that determine the physical properties of every creature. Thus, there is a good reason why DNA is called "the molecule of life," and its importance cannot be overemphasized.

The question before us is the following: How did Franklin's scientific colleagues, especially James Watson, relate to her and her research work?

After winning the Nobel Prize, James Watson decided to write a book (eventually entitled *The Double Helix*) that he called his "personal account of the discovery of the structure of DNA." In the mid-1960s, Watson obtained an agreement from Harvard University Press to publish the book that he would write. However, Watson's draft manuscript evoked severe criticism that it was needlessly hurtful in its characterization of, and offhand remarks about, many people, especially Rosalind Franklin. Therefore, Harvard University Press refused to publish it.

In response to the criticism, including criticism from Nobel Prize winners, Watson removed or watered down some of his most offensive passages. He also added an Epilogue in which he stated that his "initial impressions of Franklin, both scientific and personal (as recorded in the early pages of this book) were often wrong." He closed the Epilogue with brief posthumous praise of Franklin in an apparent effort to rectify the unfavorable picture of her that appeared in the main body of the book. However, Watson had not removed enough of the offending passages to satisfy the critics and Harvard University Press did not publish the book. It was eventually published by Atheneum, a commercial publisher.

Before relating some of Watson's derogatory statements about Rosalind Franklin, it should be noted that, if one is outraged by the slurs against Franklin that appear in *The Double Helix*, one should remember that *the original version of Watson's book was even more derogatory.*

First of all, in his book, Watson does not refer to Rosalind Franklin by her first name or last name, as he does for every other person mentioned in his book. Rather, Watson refers to her as Rosy, "as we called her from a distance." Even though he knew that she detested the name Rosy, a demeaning baby name that implies childishness and immaturity, Watson saw nothing wrong in constantly referring to Rosalind Franklin as Rosy. Only in the Epilogue, hoping to make amends, does Watson refer to her as Rosalind Franklin.

Let us now see how Watson described Franklin, his scientific colleague, in *The Double Helix*. On page 17, we read the following:

> *"By choice, Rosy did not emphasize her feminine qualities. Though her features were strong, she was not unattractive and might have been quite stunning had she taken even a mild interest in clothes. This she did not. There was never lipstick to contrast with her straight black hair. At the age of thirty-one, her dresses showed all the imagination of English bluestocking adolescents. So it was quite easy to imagine her as the product of an unsatisfied mother who unduly stressed the desirability of professional careers that could save bright girls from marriages to dull men."*

One wonders what possible relevance this degrading passage could have had in a book that claimed to describe "the discovery of the molecular structure of DNA" (Watson's subtitle).

We also find a very unflattering description of Franklin's personality (pages 16–18):

> *"Almost from the moment that she [Franklin] arrived in Maurice's [Wilkins's] lab, they began to upset each other ... Clearly, Rosy had to go or be put in her place. The former was obviously preferable because, given her belligerent moods, it would be very difficult for Maurice to maintain a dominant position that would allow him to think unhindered about DNA ... If she could only keep her emotions under control ..."*

Finally, one asks the following: Did Rosalind Franklin make a significant contribution to the discovery of the double-helix structure of DNA? Crick and Watson certainly did not think so. Nowhere in their Nobel Prize-winning article of 25 April 1953 did Crick and Watson state, or even imply, that Franklin had made a significant contribution. Franklin was only given a meaningless mention in the acknowledgments.

A similar view was expressed by the third DNA Nobelist, Maurice Wilkins. In his book about the discovery of the DNA structure, Wilkins wrote the following (*The Third Man of the Double Helix*, p. 257):

> *"Our group was of various sizes and an ever-changing composition, and there would be many talented and dedicated colleagues who came and went during those years whose contributions might be compared to Rosalind's."*

In other words, Wilkins claimed that Rosalind Franklin was just another of the talented research workers "who came and went" in his laboratory without making an especially important contribution to unraveling the structure of DNA.

A very different picture emerges from what scientists have subsequently written about Franklin's contribution to the DNA project. A number of articles published in the prestigious journals *Science* and *Nature* bear the following titles:

- "The Double Helix and the Wronged Heroine"
- "The First Lady of DNA"
- "What Rosalind Franklin Truly Contributed to the Discovery of DNA's Structure"
- "Untangling Rosalind Franklin's Role in the DNA Discovery"
- "Rosalind Franklin's Role in the DNA Discovery Gets a New Twist"

An essay in the *New York Times* issue of 25 April 2023 argued that Franklin was an "equal contributor" in the DNA discovery. Her only defect seems to have been that she was a woman who did not dress to the satisfaction of James Watson!

# Epilogue

These previous pages have been devoted to describing some of the wonders of the physical world, a subject close to my heart. The book has been restricted to physics and evolution, subjects on which I can write with a degree of confidence. I have limited myself to those topics that can be described in layman's terms, thus enabling this book to be accessible and enjoyed by anyone. No previous knowledge of science is required to understand most — but not all — of the exciting new discoveries.

The purpose of this book is to show that the physical world is far more complex than it appears to be from our personal experience. Hidden below the surface lies a fascinating world that is so wondrous that it almost defies comprehension. Some of the world's most brilliant scientists have devoted their lives to uncovering these secrets. Scientists express their findings in the jargon of their profession, but there is no reason why everyone should not share the joy of learning about these wonders.

Since scientists are human beings, even the greatest scientists sometimes make mistakes. Therefore, it seemed worthwhile to also include some of the blunders made by leading scientists, including Nobel Prize winners. These blunders do not, of course, detract from their contribution to our profound understanding of the physical world.

In a certain sense, it brings us even closer to understanding how the process of science operates and to appreciate the achievements of the scientists.

The most astounding discovery in physics is surely quantum theory. This twentieth-century theory overturned much of what we had previously thought that we understood about the physical world. Quantum theory is so strange that the great physicist Albert Einstein could not accept it.

The most astounding discovery in the science of evolution is that the animal kingdom did not evolve gradually. Here, too, the recent discovery was so unexpected that some of the leading evolutionary biologists could not accept it.

What does the future hold for us? It seems safe to say that we have not reached the end of the journey. No doubt, even greater surprises still await us.

# Author Index

# Subject Index

Printed in the USA
CPSIA information can be obtained
at www.ICGtesting.com
JSHW010045061024
70964JS00002B/17